INJECTION AND MIXING IN TURBULENT FLOW

Joseph A. Schetz
Virginia Polytechnic Institute
and State University
Blacksburg, Virginia

Volume 68
PROGRESS IN
ASTRONAUTICS AND AERONAUTICS

Martin Summerfield, Series Editor-in-Chief
New York University, New York, New York

Published by the American Institute of Aeronautics and Astronautics

American Institute of Aeronautics and Astronautics
New York, New York

Library of Congress Cataloging in Publication Data

Schetz, Joseph A.
Injection and mixing in turbulent flow.

(Progress in Astronautics and Aeronautics; 68)
Bibliography: p.
Includes index.
1. Turbulence. 2. Mixing. 3. Jets—Fluid dynamics. I. Title. II. Series.
TL507.P75 vol. 68 [TA357] 629.1′08s 79-21734
ISBN 0-915928-35-3 [620.1′064]

Table of Contents

Preface

Encountering another monograph on the fluid mechanics of turbulence, a prospective reader is entitled to inquire into the author's reasons for undertaking the project. The original motivation for writing this monograph was threefold. First, the processes of turbulent mixing of fluid streams continue to play an important role in a wide variety of practical engineering situations. These range from fuel injection in turbojet, ramjet, and piston engines, to many industrial processes in chemical and food production, and on to the dispersion of pollutants in waterways and the atmosphere, as representative examples. Second, the subject had not been treated in depth since the now classic book by Abramovich of the late 1950's. In the intervening period, whole new classes of both analytical and experimental approaches appeared. The use of the large digital computer for the numerically exact solution of the equations describing viscous flows was a rarity in 1960 and is the standard practice today. The closely related development and implementation of higher-order turbulence models, such as those based upon turbulence kinetic energy or Reynolds stress equations, also matured during this period. For experimental studies, solid-state electronics has turned the hot-wire anemometer from a troublesome instrument once used successfully only by experts into a device capable of routine use by technicians and students with training periods measured only in hours, at least for the simpler flows. Furthermore, important extensions to include quality measurements of, for example, concentration fluctuations have been completed successfully. Third, several topics under the general subject of turbulent mixing have become of much greater practical interest. These include cases with supersonic and hypersonic mainstream flows, two-phase mixtures, three-dimensional geometries, and swirl and wake-like flows where there is a zero net axial momentum defect, to name a few. This book was designed to cover as many of these developments as practical.

The organization of the volume is based upon the author's belief that the present state of knowledge in this area requires that a worker wishing to predict the development of a given mixing flow must now apply a combination of empirical and analytical techniques. Thus, each chapter and section contains, first, an overview of the available mean flow and turbulence data, and then an ordered presentation of analytical procedures based upon the complexity and presumed relative rigor of the turbulence model employed. Only methods which employ numerically exact solutions of the equations of motion are generally included. Wherever possible, competing predictions for the same flow are compared to each other and to experimental data. The case of a single jet, planar or axisymmetric, in a coflowing mainstream was selected as a baseline case and is discussed in considerable detail. The cases of axial pressure gradient, zero net axial momentum defect (self-propulsion), swirl, two-phase mixtures, three-dimensional but coaxial geometry, transverse injection, buoyancy forces, and viscous-inviscid interaction then are treated as variations on the baseline case. In order to place some bounds upon the scope of the coverage, some important but separable effects were excluded. Primary among these are the explicit influence of solid boundaries in the mixing region, chemical reactions, and noise production.

Some theoretical and experimental topics are not yet mature enough to permit coverage here in real depth at this time. The Laser Doppler Velocimeter has not been applied to enough mixing flows to have had much impact on the overall data base. Also, some controversy as to the precise nature, origin, and influence of "large-scale turbulent structures," especially at high Reynolds numbers, still exists. Lastly, the true promise of methods that attempt to treat the turbulent nature of the flow directly, including spectral methods, has yet to be tested for mixing flows of practical interest. The period of the 1980's can be expected to be one of considerable activity in these areas, and a volume such as this, prepared in another decade, would be deeply concerned with them.

This book is aimed at the engineer or physical scientist who has a background in undergraduate-level fluid dynamics and who wishes an introduction to turbulent mixing processes. It will be useful to aerospace, chemical, civil, and mechanical engineers and to chemists, geologists, meteorologists, oceanographers, and physicists who are concerned with any or all of the many situations where injection and mixing are important. The book is not really a text, but it can serve as the instructional material for a short course or a specialized graduate course.

This volume evolved from what was intended to be a review article supported by a Fellowship for that purpose awarded by The Engineering Foundation upon nomination by the American Society of Mechanical Engineers. The writer is deeply grateful for that assistance.

Several individuals and organizations in addition to The Engineering Foundation deserve special thanks. The writer was introduced to the general subject by Dr. Antonio Ferri. Though he is gone, his insights and infectious enthusiasm will remain alive in the technical community so long as any of his students or those of us who considered ourselves as such are active. The support of the Air Force Office of Scientific Research, with Dr. B. T. Wolfson as Technical Monitor, for much of the author's personal research on liquid jet injection and mixing in high-speed flows is gratefully acknowledged. The Applied Physics Laboratory of Johns Hopkins University has provided support for the majority of the writer's work on the analysis of mixing flows through a position as a Consultant. Direct collaborations with Dr. Frederick S. Billig and Mr. Stanley Favin have been stimulating, fruitful, and pleasant. Dr. S. I. Baranovsky of the Moscow Aviation Institute was very helpful with guidance through the latest Soviet literature in the field. Finally, thanks are due the authors and publishers who freely gave kind permission to include selected material from their work in this volume.

Joseph A. Schetz
July 1979

Progress in Astronautics and Aeronautics

VOLUMES	EDITORS
1. **Solid Propellant Rocket Research.** 1960	Martin Summerfield *Princeton University*
2. **Liquid Rockets and Propellants.** 1960	Loren E. Bollinger *The Ohio State University* Martin Goldsmith *The Rand Corporation* Alexis W. Lemmon Jr. *Battelle Memorial Institute*
3. **Energy Conversion for Space Power.** 1961	Nathan W. Snyder *Institute for Defense Analyses*
4. **Space Power Systems.** 1961	Nathan W. Snyder *Institute for Defense Analyses*
5. **Electrostatic Propulsion.** 1961	David B. Langmuir *Space Technology Laboratories, Inc.* Ernst Stuhlinger *NASA George C. Marshall Space Flight Center* J. M. Sellen Jr. *Space Technology Laboratories, Inc.*
6. **Detonation and Two-Phase Flow.** 1962	S. S. Penner *California Institute of Technology* F. A. Williams *Harvard University*

7. **Hypersonic Flow Research.** 1962	Frederick R. Riddell *AVCO Corporation*
8. **Guidance and Control.** 1962	Robert E. Roberson *Consultant* James S. Farrior *Lockheed Missiles and Space Company*
9. **Electric Propulsion Development.** 1963	Ernst Stuhlinger *NASA George C. Marshall Space Flight Center*
10. **Technology of Lunar Exploration.** 1963	Clifford I. Cummings and Harold R. Lawrence *Jet Propulsion Laboratory*
11. **Power Systems for Space Flight.** 1963	Morris A. Zipkin and Russell N. Edwards *General Electric Company*
12. **Ionization in High-Temperature Gases.** 1963	Kurt E. Shuler, Editor *National Bureau of Standards* John B. Fenn, Associate Editor *Princeton University*
13. **Guidance and Control—II.** 1964	Robert C. Langford *General Precision Inc.* Charles J. Mundo *Institute of Naval Studies*
14. **Celestial Mechanics and Astrodynamics.** 1964	Victor G. Szebehely *Yale University Observatory*
15. **Heterogeneous Combustion.** 1964	Hans G. Wolfhard *Institute for Defense Analyses* Irvin Glassman *Princeton University* Leon Green Jr. *Air Force Systems Command*
16. **Space Power Systems Engineering.** 1966	George C. Szego *Institute for Defense Analyses* J. Edward Taylor *TRW Inc.*

17. **Methods in Astrodynamics and Celestial Mechanics.** 1966
    Raynor L. Duncombe
    U. S. Naval Observatory
    Victor G. Szebehely
    Yale University Observatory

18. **Thermophysics and Temperature Control of Spacecraft and Entry Vehicles.** 1966
    Gerhard B. Heller
    NASA George C. Marshall Space Flight Center

19. **Communication Satellite Systems Technology.** 1966
    Richard B. Marsten
    Radio Corporation of America

20. **Thermophysics of Spacecraft and Planetary Bodies: Radiation Properties of Solids and the Electromagnetic Radiation Environment in Space.** 1967
    Gerhard B. Heller
    NASA George C. Marshall Space Flight Center

21. **Thermal Design Principles of Spacecraft and Entry Bodies.** 1969
    Jerry T. Bevans
    TRW Systems

22. **Stratospheric Circulation.** 1969
    Willis L. Webb
    Atmospheric Sciences Laboratory, White Sands, and University of Texas at El Paso

23. **Thermophysics: Applications to Thermal Design of Spacecraft.** 1970
    Jerry T. Bevans
    TRW Systems

24. **Heat Transfer and Spacecraft Thermal Control.** 1971
    John W. Lucas
    Jet Propulsion Laboratory

25. **Communication Satellites for the 70's: Technology.** 1971
    Nathaniel E. Feldman
    The Rand Corporation
    Charles M. Kelly
    The Aerospace Corporation

26. **Communication Satellites for the 70's: Systems.** 1971
    Nathaniel E. Feldman
    The Rand Corporation
    Charles M. Kelly
    The Aerospace Corporation

27. Thermospheric Circulation. 1972	Willis L. Webb *Atmospheric Sciences Laboratory, White Sands, and University of Texas at El Paso*
28. Thermal Characteristics of the Moon. 1972	John W. Lucas *Jet Propulsion Laboratory*
29. Fundamentals of Spacecraft Thermal Design. 1972	John W. Lucas *Jet Propulsion Laboratory*
30. Solar Activity Observations and Predictions. 1972	Patrick S. McIntosh and Murray Dryer *Environmental Research Laboratories, National Oceanic and Atmospheric Administration*
31. Thermal Control and Radiation. 1973	Chang-Lin Tien *University of California, Berkeley*
32. Communications Satellite Systems. 1974	P. L. Bargellini *COMSAT Laboratories*
33. Communications Satellite Technology. 1974	P. L. Bargellini *COMSAT Laboratories*
34. Instrumentation for Airbreathing Propulsion. 1974	Allen E. Fuhs *Naval Postgraduate School* Marshall Kingery *Arnold Engineering Development Center*
35. Thermophysics and Spacecraft Thermal Control. 1974	Robert G. Hering *University of Iowa*
36. Thermal Pollution Analysis. 1975	Joseph A. Schetz *Virginia Polytechnic Institute*
37. Aeroacoustics: Jet and Combustion Noise; Duct Acoustics. 1975	Henry T. Nagamatsu, Editor *General Electric Research and Development Center* Jack V. O'Keefe, Associate Editor *The Boeing Company* Ira R. Schwartz, Associate Editor *NASA Ames Research Center*

38. Aeroacoustics: Fan, STOL, and Boundary Layer Noise; Sonic Boom; Aeroacoustics Instrumentation. 1975

Henry T. Nagamatsu, Editor
General Electric Research and Development Center
Jack V. O'Keefe, Associate Editor
The Boeing Company
Ira R. Schwartz, Associate Editor
NASA Ames Research Center

39. Heat Transfer with Thermal Control Applications. 1975

M. Michael Yovanovich
University of Waterloo

40. Aerodynamics of Base Combustion. 1976

S. N. B. Murthy, Editor
Purdue University
J. R. Osborn, Associate Editor
Purdue University
A. W. Barrows and J. R. Ward, Associate Editors
Ballistics Research Laboratories

41. Communication Satellite Developments: Systems. 1976

Gilbert E. LaVean
Defense Communications Engineering Center
William G. Schmidt
CML Satellite Corporation

42. Communication Satellite Developments: Technology. 1976

William G. Schmidt
CML Satellite Corporation
Gilbert E. LaVean
Defense Communications Engineering Center

43. Aeroacoustics: Jet Noise, Combustion and Core Engine Noise. 1976

Ira R. Schwartz, Editor
NASA Ames Research Center
Henry T. Nagamatsu, Associate Editor
General Electric Research and Development Center
Warren C. Strahle, Associate Editor
Georgia Institute of Technology

44. Aeroacoustics: Fan Noise and Control; Duct Acoustics; Rotor Noise. 1976	Ira R. Schwartz, Editor *NASA Ames Research Center* Henry T. Nagamatsu, Associate Editor *General Electric Research and Development Center* Warren C. Strahle, Associate Editor *Georgia Institute of Technology*
45. Aeroacoustics: STOL Noise; Airframe and Airfoil Noise. 1976	Ira R. Schwartz, Editor *NASA Ames Research Center* Henry T. Nagamatsu, Associate Editor *General Electric Research and Development Center* Warren C. Strahle, Associate Editor *Georgia Institute of Technology*
46. Aeroacoustics: Acoustic Wave Propagation; Aircraft Noise Prediction; Aeroacoustic Instrumentation. 1976	Ira R. Schwartz, Editor *NASA Ames Research Center* Henry T. Nagamatsu, Associate Editor *General Electric Research and Development Center* Warren C. Strahle, Associate Editor *Georgia Institute of Technology*
47. Spacecraft Charging by Magnetospheric Plasmas. 1976	Alan Rosen *TRW Inc.*
48. Scientific Investigations on the Skylab Satellite. 1976	Marion I. Kent and Ernst Stuhlinger *NASA George C. Marshall Space Flight Center* Shi-Tsan Wu *The University of Alabama*
49. Radiative Transfer and Thermal Control. 1976	Allie M. Smith *ARO Inc.*

50. Exploration of the Outer Solar System. 1977	Eugene W. Greenstadt *TRW Inc.* Murray Dryer *National Oceanic and Atmospheric Administration* Devrie S. Intriligator *University of Southern California*
51. Rarefied Gas Dynamics, Parts I and II (two volumes). 1977	J. Leith Potter *ARO Inc.*
52. Materials Sciences in Space with Application to Space Processing. 1977	Leo Steg *General Electric Company*
53. Experimental Diagnostics in Gas Phase Combustion Systems. 1977	Ben T. Zinn, Editor *Georgia Institute of Technology* Craig T. Bowman, Associate Editor *Stanford University* Daniel L. Hartley, Associate Editor *Sandia Laboratories* Edward W. Price, Associate Editor *Georgia Institute of Technology* James G. Skifstad, Associate Editor *Purdue University*
54. Satellite Communications: Future Systems. 1977	David Jarett *TRW Inc.*
55. Satellite Communications: Advanced Technologies. 1977	David Jarett *TRW Inc.*
56. Thermophysics of Spacecraft and Outer Planet Entry Probes. 1977	Allie M. Smith *ARO Inc.*
57. Space-Based Manufacturing from Nonterrestrial Materials. 1977	Gerard K. O'Neill, Editor *Princeton University* Brian O'Leary, Assistant Editor *Princeton University*

58. Turbulent Combustion. 1978	Lawrence A. Kennedy *State University of New York at Buffalo*
59. Aerodynamic Heating and Thermal Protection Systems. 1978	Leroy S. Fletcher *University of Virginia*
60. Heat Transfer and Thermal Control Systems. 1978	Leroy S. Fletcher *University of Virginia*
61. Radiation Energy Conversion in Space. 1978	Kenneth W. Billman *NASA Ames Research Center*
62. Alternative Hydrocarbon Fuels: Combustion and Chemical Kinetics. 1978	Craig T. Bowman *Stanford University* Jørgen Birkeland *Department of Energy*
63. Experimental Diagnostics in Combustion of Solids. 1978	Thomas L. Boggs *Naval Weapons Center* Ben T. Zinn *Georgia Institute of Technology*
64. Outer Planet Entry Heating and Thermal Protection. 1979	Raymond Viskanta *Purdue University*
65. Thermophysics and Thermal Control. 1979	Raymond Viskanta *Purdue University*
66. Interior Ballistics of Guns. 1979	Herman Krier *University of Illinois at Urbana-Champaign* Martin Summerfield *New York University*
67. Remote Sensing of Earth from Space: Role of "Smart Sensors." 1979	Roger A. Breckenridge *NASA Langley Research Center*
68. Injection and Mixing in Turbulent Flow. 1980	Joseph A. Schetz *Virginia Polytechnic Institute and State University*

(Other volumes are planned.)

Nomenclature

a_1, a_D, a_s	= proportionality constants
A	= area
$b, b_{1/2}$	= mixing zone width, planar half-width
C_j, C_1, C_{g1}, etc.	= proportionality constants
C_D	= drag coefficient or proportionality constant
C	= mean concentration
C^*	$= (\rho_a - \rho_c) / (\rho_a - \rho_j)$
c	= fluctuating concentration
C_d	= discharge coefficient
C_p	= specific heat
C_t, C_t'	= Craya-Curtet numbers
$C(r)$	= scale of segregation
d, d_j	= jet diameter or minor axis for three-dimensional (3-D) orifice
d_{eq}	= equivalent diameter of 3-D injector
d_f	= frontal dimension of 3-D injector
d_s	= streamwise dimension of 3-D injector
D	= diffusion coefficient or duct or body diameter
D_T	= turbulent diffusion coefficient
e	= entrainment
E	$= e / (\rho \Delta U) L$
e^*	= eccentricity of 3-D orifice
$E(K)$	= 3-D energy spectrum function
$E_s(K)$	= integrated scalar spectrum function
$f(\ )$	= similarity profile shape function
$f(r)$	= defined in Eq. (34)
F	= defined in Eq. (86)
Fr	= Froude number, Eq. (178)
$F_1(K), F_2(K), F_3(K)$	= one-dimensional energy spectrum function
g	$= \overline{c^2}$
$g(\ )$	= particle mass flux
$g(r)$	= defined in Eq. (35)
G	= defined in Eq. (100)
Gr	= Grashof number, Eq. (174)
H	= defined in Eq. (117)
h	= half-height of planar jet or penetration
h_2	= half-height of planar duct
I_1, I_2, I_3, I_4	= profile shape integrals from Ref. 92
$\mathcal{I}_1, \mathcal{I}_2, \mathcal{I}_3$	= defined in Eqs. (81) and (137)
J	= momentum flux

I_s	= intensity of segregation
j	= 1 for axisymmetric, = 0 for planar flow
k	= turbulent kinetic energy
K	= wave number
k_1, k_2	= proportionality constants
Le_T	= turbulent Lewis number
l^0	= length of the recirculation zone
l, l_m, l^*, l'_m	= length scales in turbulence models
L_ϵ	= dissipation scale
m	$= U_\infty / U_j$
n	$= \rho_\infty / \rho_j$
M	= Mach number or parameter defined in Eq. (117)
$\mathfrak{M}$	= defined in Eq. (116)
$\dot{m}$	= mass flow in the mixing zone
N	= peripheral length or concentration flux
P	= mean pressure
p	= fluctuating pressure
Pr	= Prandtl number
Pr_T	= turbulent Prandtl number
P_{eb}	= effective back pressure
P_V, P_T, P_C	= power decay law exponents
$\mathcal{P}$	= production of k
$\mathcal{P}_{ij}$	= production of $\overline{u_i u_j}$
$\mathcal{P}_{iT}$	= production of $\overline{u_i T'}$
$\overline{\mathcal{P}}$	= defined in Eq. (89)
$\bar{q}$	$= \rho_j U_j^2 / \rho_\infty U_\infty^2$
Q	= total mass flow, Eq. (115)
r	= radial coordinate
$r_{1/2}$	= half-radius
r_j	= jet radius
Re	= Reynolds number
Re_T	= turbulent Reynolds number, $= (\quad U) b / \nu_T$
R_0	= duct radius
S_z	= source term in Eq. (20)
Sc_T	= turbulent Schmidt number
S	= swirl number
S^*	= defined in Eq. (181)
S_x	= local swirl number
T	= mean temperature
T'	= fluctuating temperature
$T_s(K)$	= transfer function
T^*	= angular momentum
T_{bv}	= Brunt-Vaisala period, Eq. (182)
u	= fluctuating axial velocity
U	= mean axial velocity
U_{av}	= average velocity in mixing zone
U_e	$= U_\infty / (U_c - U_\infty)$
$U_\delta(x)$	= secondary stream velocity

v	= fluctuating normal velocity
V	= mean normal velocity
w	= fluctuating transverse velocity
W	= mean transverse velocity and k/l^2
W_0	= W_{max}/U_{av}
x	= axial coordinate
x_c	= initial region length
x_0	= virtual jet origin shaft
y	= normal coordinate
$y_{1/2}$	= half-width in y direction
z	= transverse coordinate
$z_{1/2}$	= half-width in z direction
Z	= $k^{\alpha} l^{\beta}$
α	= exponent of k in Z
β	= exponent of l in Z
$\Delta\zeta$	= width in the x-y plane
Δz	= width perpendicular to the x-y plane
ΔU	= $U_{max} - U_{min}$
ρ	= mean density
ρ'	= fluctuating density
ν	= kinematic viscosity
ν_T	= turbulent (eddy) kinematic viscosity
μ_T	= turbulent viscosity
κ	= constant in Eq. (13)
κ_T	= turbulent thermal conductivity
κ_m	= molecular thermal conductivity
τ_T	= turbulent shear
σ_k	= Prandtl number for k
σ_g	= Prandtl number for g
σ_Z	= Prandtl number for Z
$\sigma_{r\theta}$	= defined in Eq. (132)
σ_W	= Prandtl number for W
λ	= $\rho_j U_j / \rho_\infty U_\infty$
λ_g, λ_f	= turbulent microscales
λ_s	= defined in Eq. (131)
$\Lambda_g, \Lambda_f, \Lambda_c$	= integral scales
ϵ	= dissipation
ϵ_p	= 1.0 minus volume fraction of particles
δ^*, Δ^*	= displacement thickness
θ	= momentum thickness, injection angle, or local flow angle
γ	= intermittency
ψ	= stream function
ξ	= arc length along trajectory and transformed axial coordinate
Φ	= W_{max}/U_{max}
η	= transverse coordinate (see Fig. 144)

Subscripts

a = ambient conditions
c = centerline (axis) conditions
j = initial injectant conditions
∞ = freestream conditions

I. INTRODUCTION

A. Background

The processes of injection and mixing find wide application throughout engineering and science. The combinations of flow rate, geometrical size, and thermophysical properties of practical interest are almost always such as to produce turbulent flow. Therefore, experimental and analytical studies of injection and mixing in turbulent flow go back many years, and the literature in the field is very rich. The rate of work in this area has, however, not diminished; indeed it seems to be accelerating, with journal articles and technical reports appearing at a rapid rate.

Engineering design requirements in this field can often conspire to result in flow problems that are very complex from an analysis point of view. Perhaps two concrete examples can best serve to illustrate that point as well as to provide some notion of the scope of the general problem area. Consider first the design of a condenser cooling water intake and discharge system for a thermal power station. These plants are frequently located on waterways, so that sufficient water is readily available. In recent years, however, strict limitations have been imposed for environmental reasons. Accordingly, the designer is typically confronted with rules that limit the size of the heated discharge water "mixing zone" to no more than half the width of the river and no more than half the total flow cross-sectional area. Violation of these rules can result in temporary or permanent shutdown of the generating station, and so the requirements for precision in preconstruction analysis are quite stringent. The actual flow problem involved can be characterized as fully three-dimensional with irregular boundaries involving buoyancy forces and mainstream flow rates varying by an order of magnitude, sometimes at a rate such that transient phenomena are important. Also, massive recirculation zones have been found to exist under some combinations of conditions where heated discharge water actually is drawn into the upstream cooling water intake. This can result in significant losses in overall operating efficiency. There are simply no available analyses that can provide solutions for such a problem with reliably high precision.

A second actual engineering application is the design of a liquid fuel injection system for a supersonic combustion ramjet engine. Here, one is concerned with providing a nominally uniform coverage of fuel over a typically circular combustor cross section. For fuel/air ratios of interest with hydrocarbon fuels, the periphery of the combustion chamber would likely contain a series of small-diameter injectors, and insuring sufficient fuel jet penetration into the high-speed airstream can be a severe problem. This flow problem can also be characterized as fully three-dimensional, and now we are also concerned with liquid jet breakup and atomization leading to two-phase mixing of the resulting droplets with the air. The injection process itself produces complicated shock patterns in the flow, and various other compressibility effects are clearly also important. Lastly, ignition and combustion processes are superimposed on the atomization and mixing activities. Again, there are no reliably accurate analyses available to treat a complete problem of this type.

B. Intent of This Volume

How then can the engineer proceed to analyze the kinds of problems just described in order to make informed design decisions? Today, he must rely upon a combination of empirical information and simplified analyses of various component parts of a complex problem. That procedure suggests the philosophy of this volume. We shall begin with the simplest injection and mixing problem of general engineering interest, a jet in a coflowing stream, and then add complicating effects one by one. These include 1) axial pressure gradients, 2) momentumless (net) conditions, 3) swirl, 4) two-phase mixing, 5) three-dimensional, coaxial geometries, 6) density stratification, 7) transverse injection, and 8) supersonic "base" flows. Chemical reactions in turbulent flow have been specifically excluded in order to place some bounds upon the coverage. This is a large and rapidly developing field even in the absence of mixing processes, and such material is best treated in a separate volume. The same can be said for noise generation by shear flows. Indeed, other volumes in this series are devoted to that subject; it will not be treated here. Finally, the treatment of mixing flows near solid surfaces and the direct influences of walls, in general, are not covered. Thus, only what are often called "free turbulent shear flows" are included. In each section, both mean-flow and turbulence data are presented, followed by the existing analytical approaches. The volume closes with a brief survey of the major experimental techniques that are used in this field of study.

A second facet of the philosophy of this volume is concerned with the depth of the coverage in addition to the scope that was just discussed. First, the material herein is not aimed at the expert in any or all of the subjects covered. By definition, a review cannot hope to provide new information to one who is familiar with the literature reviewed. It is written for the layman with a solid general fluid dynamics background who wishes to become familiar with this problem area or any parts of it. Second, historical and geographical completeness were not taken as important criteria for the selection of the specific items included. In general, the latest and/or most thorough studies were used, even though prior work in the same area may represent a greater scientific accomplishment. Third, the writer's own experience and interests obviously color the selection and the presentation of the material. This is not intended as an apology but rather as a clear notice to the reader.

It was stated earlier that the literature in the field is rich; indeed, there have been a number of books and review articles published over the years. Where then does the coverage and philosophy of the current volume fit? That question is answered most easily and efficiently by reference to Tables 1 and 2, which summarize the coverage of the various major previous publications.[1-24] It should be clear that this work has sought to provide a much broader

Table 1 Coverage of experimental data in books and review articles on injection and mixing in turbulent flow(where M = mean-flow data, T = turbulence data)

Author	Date	Jet $U_\infty=0$		Jet $U_\infty\neq0$		Var. T		Var. C		Press. Grad.		Zero Mom. D.		Swirl		Two Ph.		3 D		Plume		Strat.		Ang. Inj.		Remarks
		M	T	M	T	M	T	M	T	M	T	M	T	M	T	M	T	M	T	M	T	M	T	M	T	
Schlichting	1942*	X				X																				
Pai	1954	X	X			X		X																		
Hinze	1959	X	X			X	X																			
Abramovich	1960	X	X	X	X	X	X	X		X						X				X				X		
Ferri	1965			X		X		X		X																
Ferri	1967			X		X		X		X						X										
Newman	1967			X						X																
NASA	1970	X	X	X	X	X	X	X																		
Bradbury	1971	X	X	X	X																					
Tennekes & Lumley	1972																									Good Ref. No mix. data
Launder & Spalding	1972	X	X	X		X		X	X																	
Murthy	1974																									Gen. Disc. No data
Abramovich	1974	X	X	X	X	X								X												
Abramovich	1975	X		X		X		X						X		X	X									
Rodi	1975	X	X	X	X					X	X															
Rajaratham	1976	X	X	X	X					X	X			X	X					X				X		
Bradshaw(edit.)	1976			X						X	X															
Townsend	1976**	X	X																							
Present Volume	1979	X	X	X	X	X	X	X	X	X	X	X	X	X	X	X	X	X		X	X	X	X	X	X	

* First German Edition ** Latest Edition

Table 2 Coverage of turbulent analyses in books and review articles on injection and mixing in turbulent flow

Author	Date	Mean - Flow Models											One - Equation Models										
		Jets	Var. T	Var. C	Press. Grad.	Zero Mom. D.	Swirl	Two Ph.	3D	Plume	Strat.	Ang. Inj.	Jets	Var. T	Var. C	Press. Grad.	Zero Mom. D.	Swirl	Two Ph.	3D	Plume	Strat.*	Ang. Inj.
Schlichting	1942*	X	X																				
Pai	1954	X	X	X																			
Birkhoff & Zarantonello	1957	X				X		X															
Hinze	1959	X	X	X				X															
Abramovich	1960	X	X	X	X		X	X		X		X											
Ferri	1965	X	X	X	X																		
Brodkey	1966	No Solutions																					
Ferri	1967	X	X	X	X																		
Newman	1967	X			X																		
NASA	1970	X	X	X									X	X	X								
Bradbury	1971	X																					
Harsha	1971	X	X	X									X	X	X								
Tennekes & Lumley	1972	X	X			X				X													
Launder & Spalding	1972	X	X	X									No solutions for jets										
Murthy	1974	No Solutions											No solutions										
Abramovich	1974	X	X	X				X					X										
Abramovich	1975	X	X	X				X	X				X										
Bradshaw	1972,75	No Solutions																					
Rodi	1975	Similarity Solutions Only																					
Rajaratham	1976	X			X		X					X											
Bradshaw(edit.)	1976	X	X	X	X			X															
Townsend	1976**	X			X																		
Present Volume	1979	X	X	X	X	X	X	X	X	X	X	X	X	X	X	X	X	X			X	X	X

Table 2 (continued)

		Two - Equation Models											Reynolds Stress Models										
		Jets	Var. T	Var. C	Press. Grad.	Zero Mom. D.	Swirl	Two Ph.	3D	Plume	Strat.	Ang. Inj.	Jets	Var. T	Var. C	Press. Grad.	Zero Mom. D.	Swirl	Two Ph.	3D	Plume	Strat.	Ang. Inj
Schlichting	1942*																						
Pai	1954																						
Birkhoff & Zarantonello	1957																						
Hinze	1959																						
Abramovich	1960																						
Ferri	1965																						
Brodkey	1966																						
Ferri	1967																						
Newman	1967																						
NASA	1970	X	X	X									X										
Bradbury	1971	X																					
Harsha	1971																						
Tennekes & Lumley	1972																						
Launder & Spalding	1972	X											X		X							X	
Murthy	1974																						
Abramovich	1974																						
Abramovich	1975																						
Bradshaw	1972,75																						
Rodi	1975																						
Rajaratham	1976																						
Bradshaw(edit.)	1976																						
Townsend	1976**																						
Present Volume	1979	X	X	X	X		X		X	X			X	X	X		X	X				X	

* First German Edition ** Latest Edition

coverage than before as well as to discuss the major new developments that have occurred just recently.

C. Characterization of Analyses of Free Turbulent Flows

Until the mid-1960's there were two major difficulties to be overcome in producing a useful analysis of free turbulent shear flow problems: modeling of the turbulent transport processes, and solution of the resulting system of differential equations. Often these two matters became intimately connected. For example, mathematical difficulties in solving the equations clearly influence the complexity of the turbulence model chosen, since this directly affects the difficulty of the mathematical problem. Furthermore, crudities in the predictions introduced by necessary approximations used in obtaining solutions to the differential equations confuse the comparison with experiment and the judgment of the adequacy of a given turbulence model. Fortunately, the advent of the widespread use of large digital computers has all but eliminated the problem of solving the differential equations for many problems of practical interest. Numerically "exact" solutions can be readily obtained for the systems of equations that result from employing turbulence models of quite considerable complexity with only modest restrictions and idealizations on problem geometry. Of course, some workers have capitalized on this advancement to try to treat very complicated physical cases with a minimum of idealization. Others have tried to implement new approaches to the turbulence modeling question which also tax the limits of current computing machines. This situation is likely to persist for the foreseeable future. Practicing engineers will be using the best available models that can be run on available machines at reasonable cost, whereas workers in the field will be constantly pushing the limits of the machines available to them at that instant in time.

There is a choice as to the numerical method to be employed to obtain a numerical solution for a given problem using the modern, large-scale digital computer. Here, we shall give only a brief overview of two of the available classes of methods using the one-dimensional, unsteady heat equation as a simple example. This equation has been found to serve well as a linearized, "model" equation to represent the boundary-layer equations for numerical analysis studies. The differential equation to be considered here is then

$$\frac{\partial T}{\partial t} = \frac{\partial^2 T}{\partial x^2} \tag{1}$$

with initial conditions $T=f(x)$ when $t=0$ and boundary conditions $T(0,t) = T(\pi,t) = 0$ for $t \geq 0$.

Up to the present time, by far the most common approaches are based on the method of finite differences. In that case, the differentials are replaced by finite differences as, for example,

$$\frac{\partial T}{\partial t} \approx \frac{T_{m,n+1} - T_{m,n}}{\Delta t} \tag{2a}$$

$$\frac{\partial^2 T}{\partial x^2} \approx \frac{T_{m+1,n} - 2T_{m,n} + T_{m-1,n}}{(\Delta x)^2} \tag{2b}$$

Here, m and n denote locations in a rectangular grid formed by dividing up the region of interest by Δx and Δt. The solution for the dependent variable (T, temperature in this example) at each grid point is then found from a system of algebraic equations:

$$T_{m,n+1} = \frac{\Delta t}{(\Delta x)^2}\,[T_{m+1,n} + T_{m-1,n}] - \left[\frac{2(\Delta t)}{(\Delta x)^2} - 1\right] T_{m,n} \tag{3}$$

Such systems of equations are easily and quickly solved on the digital computer for nearly arbitrary initial functions, $f(x)$.

The second class of methods which is gaining popularity, especially for the attempts at direct numerical simulation of turbulence, is called spectral methods. The exact solution to the simple problem given by Eq. (1) can be obtained by Fourier series as

$$T(x,t) = \sum_{N=1}^{\infty} A_N(t)\sin(Nx) \tag{4}$$

The partial differential equation reduces to

$$\frac{dA_N}{dt} = -N^2 A_N \tag{5}$$

and

$$A_N(0) = F_N, \qquad f(x) = \sum_{N=1}^{\infty} A_N \sin(Nx)$$

For the spectral method, the infinite sums in the preceding exact expressions are replaced by finite sums, i.e.,

$$T_{\bar{N}}(x,t) = \sum_{N=1}^{\bar{N}} A_N(t)\sin(Nx) \tag{6}$$

$$\frac{dA_N}{dt} = -N^2 A_N; \qquad N = 1,2,\ldots,\bar{N}, \qquad A_N(0) = F_N$$

The utility of this approach is not immediately obvious here but is a result of the fact that this method can be efficiently implemented on a large digital computer using the method of Fast Fourier transforms.

The spectral methods have only begun to be applied in practical flow configurations. Thus, all of the results of computations to be presented in this volume will be based upon finite-difference formulations.

The result of this dramatic, sudden advance in computational capability has been that most attention in the past decade or so has been on the "exact" implementation of older turbulence models and the development of some altogether new models. We have been using the term "model" here frequently, and we can now proceed to make the meaning of the term clear. This is conveniently done while describing the current hierarchy of models which will serve as background for the later detailed considerations of the application of each type of model to the problems of interest. It is easiest to discuss the equations for a reasonably simple situation for clarity. The idealizations to be introduced here are not to be taken as basic to the rest of this volume. If we assume a steady state in a planar geometry with a homogeneous, low-speed, constant-temperature flow and further restrict ourselves to "thin" shear layers, so that the boundary-layer approximations apply, the lowest-order system of equations that can be used may be written as:

$$\frac{\partial U}{\partial x} + \frac{\partial V}{\partial y} = 0 \tag{7}$$

$$U\frac{\partial U}{\partial x} + V\frac{\partial U}{\partial y} = -\frac{1}{\rho}\frac{\mathrm{d}P}{\mathrm{d}x} + \frac{\partial}{\partial y}(-\overline{uv}) \tag{8}$$

These are mean flow equations in that they are time-averaged, and capitalized dependent variables are means, whereas lower-case dependent variables are fluctuating quantities. A bar over any fluctuating quantity or combination of them denotes a time average also. This can be illustrated using the streamwise velocity as an example:

$$u^*(x,y,z,t) = U(x,y,z) + u(x,y,z,t) \tag{9}$$

with

$$U(x,y) = \frac{1}{T}\int_0^T u^*(x,y,z,t)\,\mathrm{d}t \tag{10}$$

and T is a time long compared to a characteristic fluctuation time. It is well to note here that, in the real world, there is no such thing as two-dimensional turbulence even for a flow that is two dimensional in the mean.

For a boundary-layer problem specified by Eqs. (7) and (8), the pressure field $P(x)$ is taken as imposed upon the shear flow by an external, inviscid flow, so that we would expect to solve these two equations for the two unknowns (U, V). However, we notice an additional unknown term as the last on the right-hand side of Eq. (8). This is a momentum transfer term that plays the same role as simple Newtonian shear in a laminar flow, so that the grouping $(-\rho\overline{uv}) \equiv \tau_T$ is termed the turbulent shear or Reynolds stress. Working within the framework of Eqs. (7) and (8), it is necessary to relate this term to the other independent or dependent variables in order to close the problem from a mathematical standpoint. Since these other variables are all means, such a model is termed a "mean-flow model." An example of a model of that kind would be the mixing length model (see Ref. 1):

$$\tau_T = -\rho\overline{uv} = \rho l_m^2 \left|\frac{\partial U}{\partial y}\right| \frac{\partial U}{\partial y} \tag{11}$$

where l_m is a mixing length which is usually related to the width of the shear zone $b(x)$ in a simple way, e.g., $l_m = \text{const} \times b(x)$ with the constant $\approx 1/10$. This model is very closely related to the eddy viscosity model:

$$\tau_T = -\rho\overline{uv} = \rho\nu_T\frac{\partial U}{\partial y} \tag{12}$$

Here ν_T is an effective eddy kinematic viscosity in analogy with the usual laminar kinematic viscosity ν. However, since ν_T is not a property of the fluid itself, but rather of the flowfield, it must still be related to the rest of the variables or other characteristics of the flow. A widely used example is[1]

$$\nu_T = \kappa b(x) \mid U_{\max} - U_{\min} \mid \tag{13}$$

where κ is an empirical constant $\approx 3/100$.

Several observations are in order. First, looking at Eqs. (11) and (12), we see that

$$\nu_T = l_m^2 \left| \frac{\partial U}{\partial y} \right| \tag{14}$$

so that there is essentially no fundamental difference between these two models. Second, both Eqs. (11) and (12), by themselves, involve a severe restriction without actually resolving the difficulty. Both are restricted to a gradient transport formulation:

$$\tau_T \sim \frac{\partial U}{\partial y} \tag{15}$$

and this condition is known to be violated in some turbulent flows. Problems most often arise where the mean-flow velocity profile contains one or more inflection points. It still remains to make a further statement relating either l_m or ν_T to the flowfield. Lastly, neither these two nor any other mean-flow model directly treats any aspect of the fluctuating character of the turbulent flow itself.

The empirical "constants" involved in the mixing length and eddy viscosity models have been found to be a function of some features of the turbulence field. It is possible to use empirical information to generate a simple algebraic expression, so that the "constants" become functions of some simple turbulence quantity. This approach may be termed an "algebraic turbulence function model."

The first turbulent transport model that dealt directly with the actual turbulent nature of the flow was proposed by Prandtl and Kolmogorov in the 1940's (see Ref. 15 for a clear exposition). They chose, as a simple representative turbulence quantity, the kinetic energy of the fluctuations defined as

$$k^2 = \overline{(u^2 + v^2 + w^2)}/2 \tag{16}$$

The variation for k throughout the flow is determined via the solution of a separate transport equation for this quantity. This equation is derived from the Navier-Stokes equations by multiplying each component equation by the corresponding component of the fluctuating velocity, time averaging, summing all three equations, and then, usually, making simplifying assumptions such as with the boundary-layer approximation. Under the same restrictions as for Eqs. (7) and (8), the result is

$$\underset{\text{convection}}{\rho\frac{\mathrm{D}k}{\mathrm{D}t}} = \underset{\text{diffusion}}{-\frac{\partial}{\partial y}(\rho\overline{vk} + \overline{vp})} \underset{\text{production}}{-\rho\overline{uv}\frac{\partial U}{\partial y}} \underset{\text{viscous dissipation}}{-\mu\sum\overline{\left(\frac{\partial u_i}{\partial x_j}\right)^2}} \tag{17}$$

This type of model is usually appropriately termed a "one-equation model." It is sometimes also called a first-order model. It is important to mention here that further modeling is still necessary in order to use this approach to close the whole system. One also needs to solve Eqs. (7) and (8) or their equivalent, and τ_T must be related to k. This is usually done by reinvoking the eddy viscosity concept, but now with

$$\nu_T = \rho\sqrt{k}\, l \tag{18}$$

We retain $l = \text{const} \times b$. Note that the gradient transport concept with its limitations remains. More recently, and mostly for wall-dominated flows, Bradshaw[25] has suggested the use of a direct relationship

$$\tau_T = a_l \rho k \tag{19}$$

where $a_l \approx 0.30$. This model does avoid the gradient transport concept, but it is not generally felt to be as widely applicable as Eq. (18) for free shear flows.

There are a number of situations of practical interest where a suitable value of l cannot be found from a simple algebraic expression such as used in the foregoing. In the present context of mixing problems, a classic example is the case of the merging of the mixing zones from several parallel, coaxial jets, all exhausting in one cross-sectional plane. Near the point of injection, each jet is unaware of the others, and the mixing length can be simply related to the local width of a single jet's mixing zone. Further downstream, the several mixing zones have grown to the point where they merge, more or less abruptly. The "width" of the mixing zone has now suddenly grown much larger. If there are 50 jets, it becomes 50 times larger. Clearly, the local flow in one of the middle jets cannot instantly respond to the presence of all of these other jets in a linear way, as would be implied by taking l as proportional to the new mixing zone width. We must have something better, and one approach has been to seek an independent equation for l or for a new quantity that is a combination of k with l, $Z \equiv k^\alpha l^\beta$. Such an equation can also be found by manipulating the Navier-Stokes equations, and the result under the same restrictions as before and following some modeling is

$$\rho \frac{DZ}{Dt} = \frac{\partial}{\partial y}\left(\frac{\mu_T}{\sigma_z}\frac{\partial Z}{\partial y}\right) + Z\left[C_l \frac{\mu_T}{k}\left(\frac{\partial U}{\partial y}\right)^2 - C_2 \frac{\rho^2 k}{\mu_T}\right] + S_z \tag{20}$$

Here σ_z is a "Prandtl number" for diffusion of Z, S_z are secondary source terms that appear in some models, C_l and C_2 are constants, and $\mu_T = \rho\nu_T$. This equation is almost identical in form to some of the modeled forms of Eq. (17). At least a dozen workers have been active in this area. One still works with the mean-flow equations and the turbulent kinetic energy (TKE) equation along with the new Z equation, and this approach has been named a "two-equation model." Again, Ref. 15 contains a readable description.

The one-equation and two-equation models described previously both generally suffer from the gradient transport concept restriction. The eddy viscosity (or mixing length) notions combined with the gradient transport relationship, in general, imply too direct a connection between the mean flowfield and the turbulent stress. This remains a limitation on the models that are more advanced than the mean-flow models, even though other facets of the turbulence have been included. One way around this problem is to formulate a transport equation for the Reynolds stress, $-\rho\overline{uv}$, itself. With our usual restrictions for this section, the result is [15]

$$\underset{\text{convection}}{\frac{\mathrm{D}}{\mathrm{D}t}(\overline{uv})} = \underset{\text{production}}{-\overline{v^2}\frac{\partial U}{\partial y}} - \underset{\text{diffusion}}{\frac{\partial}{\partial y}\left(\overline{uv^2} + \frac{\overline{pu}}{\rho}\right)} + \underset{\text{redistribution}}{\overline{\frac{p}{\rho}\left(\frac{\partial u}{\partial y} + \frac{\partial v}{\partial x}\right)}} - \underset{\text{viscous dissipation}}{2\nu\sum\left(\overline{\frac{\partial u}{\partial x_j}\frac{\partial v}{\partial x_j}}\right)} \tag{21}$$

Again, this equation contains unknown terms that must be modeled, and different approaches have been tried. Equations like (7) and (8) are still needed, and a typical formulation might also use separate equations for l and $\overline{u^2}$, $\overline{v^2}$, $\overline{w^2}$ as well as $\overline{vw}$, $\overline{uw}$. These models might best be termed "Reynolds stress models," although they have been called "multi-equation models" for obvious reasons. They are also named "second-order models," since second order correlations are treated directly.

Up to this point, we have not made any connection with the large body of knowledge that has grown in what might be called "statistical" turbulence research. Only recently, however, have tentative connections between the statistical theories and analysis needs of the engineer been made. A general discussion of the matter as of a few years ago is given in Ref. 26. Here, we shall call such models "direct turbulence models." The basic problem with attempting to treat the fluctuating turbulent flow directly is that such flows are characterized by a range of the excited scales of motion over several orders of magnitude. Even the most optimistic projections for the capacity of future computing machines fall far short of that estimated to meet this need. Only at low Reynolds number conditions does all-scale resolution of the turbulence seem likely in the foreseeable future. One promising method for alleviating this problem which has been proposed is to limit the attempt at direct, three-dimensional, unsteady treatment of the turbulence to only those scales above a certain size. All of the scales that cannot be resolved are then modeled as a "subgrid turbulence" using an eddy viscosity or other transport approximation. The hope in this approach is that the smaller-scale structure of the turbulence is nearly universal, so that accurate resolution is not required. The large eddy structure that is presumed to contain the part of the turbulence that changes markedly from flow to flow or condition to condition is treated directly. Another direct treatment is the "vortex-dynamics" approach, where a turbulent viscous zone is modeled by many individual, inviscid vortices that are tracked with time. An attempt is made to simulate the action of molecular viscosity by adding a random walk component to each vortex each time step.

There is one other general type of turbulent mixing analysis that has developed, mostly amongst chemical engineers (see Ref. 7). It too attempts to use the results of the statistical theory of turbulence but in a quite approximate fashion. It is restricted to the latter stages of a mixing process and has mainly been applied to flows in pipes and vessels. If we consider two fluids of concentration fraction A and B with fluctuating values a and b and define

$$a' \equiv \sqrt{\overline{a^2}} \tag{22}$$

an "intensity of segregation" and a "scale of segregation" can be introduced via a concentration correlation:

$$C(r) = \frac{\overline{a(x) \cdot a(x+r)}}{(a'^2)} \tag{23}$$

The scale of segregation is

$$L_s \equiv \int C(r) \mathrm{d}r \tag{24}$$

and the intensity of segregation is

$$I_s \equiv (a'^2)/A \cdot B = (a')^2/(a_0')^2 \tag{25}$$

which varies from unity down to zero for complete mixing. The equations for the mean and fluctuating values of the concentration are written as

$$U\frac{\partial A}{\partial x} + \frac{\partial}{\partial y}(\overline{va}) = D\frac{\partial^2 A}{\partial y^2} \tag{26}$$

$$U\frac{\partial (a')^2}{\partial x} + 2\overline{va}\frac{\partial A}{\partial y} + \frac{\partial}{\partial y}(\overline{a^2 v}) = D\frac{\partial^2 (a')^2}{\partial y^2} - 2D\overline{\left(\frac{\partial a}{\partial x_i}\right)\left(\frac{\partial a}{\partial x_j}\right)} \tag{27}$$

where D is the molecular diffusivity. This system must be supplemented with an additional "turbulence" equation. With the assumption of "local homogeneity," the spectral equation

$$U\frac{\partial E_s(K)}{\partial x} - \frac{K}{3\sqrt{3}}\frac{\partial E_s(K)}{\partial x}\frac{\partial U}{\partial y} - T_s(K) = \frac{D}{2}\frac{\partial^2 E_s(K)}{\partial x_i \partial x_i} - 2DK^2 E_s(K) \tag{28}$$

has been used. Here $E_s(K)$ is the integrated scalar spectrum function, and K is the wave number:

$$(a')^2 = \int_0^\infty E_s(K)\mathrm{d}K \tag{29}$$

and thus

$$I_s = \frac{1}{(a_0')^2}\int_0^\infty E_s(K)\mathrm{d}K \tag{30}$$

and $T_s(K)$ is a transfer function that must be modeled. This is often accomplished with an eddy viscosity model.

We have now presented a very brief overview of six levels of turbulent transport models, and the reader might very well ask which he should try to use for a given type of flow problem. Unfortunately, the answer to that reasonable question is not all clear at the present time. Generally, the higher the level of the model, the better will be the agreement with experiment, although the improvement is often small. This improvement may also have to be purchased at a dear price in terms of added complexity and computational cost. For the injection and mixing problems of interest here, we shall try to provide an essentially complete coverage of the applications of the different levels of models and comparisons with experiment, although not all of the models have been applied to all of the situations of importance in this field.

Regretfully, the key information of computational time is often either not reported at all or incompletely given by the workers who have used the various approaches. The reader can, in any event, try to make his own judgement of the relative worth of the various models.

II. PARALLEL JET IN A MOVING STREAM

A. Introduction

The case of a two-dimensional (either planar or axisymmetric) jet in a moving stream of constant velocity has been selected as the baseline case to which the influence of other complicating factors will be referenced. The limiting case with an external velocity equal to zero was not chosen, since it turns out to be an extremely simple limiting case, and it is not often of interest in practice. If the velocity of the external stream changes with distance, then a pressure gradient is implied through Bernoulli's equation, and that factor is treated in a later chapter. Separate chapters are also devoted to cases with a zero net momentum defect, swirl, two-phase flow, three-dimensional flows, density stratification, injection at an angle to the main flow, and injection in the base behind bodies moving at supersonic speeds.

The first sections of this chapter describe the current state of experimental knowledge for this flow problem. We begin with the case of homogenous composition and constant temperature at low-speed conditions, followed by subsections on variable temperature and composition. In each case, both mean-flow and turbulence data are given. The experimental part of the chapter closes with a brief discussion of "large-scale structures" in jet mixing problems. The final part of the chapter contains sections that describe the various analyses of these problems and comparisons with data.

B. Experimental Information

1. Initial Region

The general flowfield of interest here is depicted schematically in Fig. 1. In the idealized case of small boundary layers both on the outside and the inside of the jet injector, there is a substantial initial region before the developing

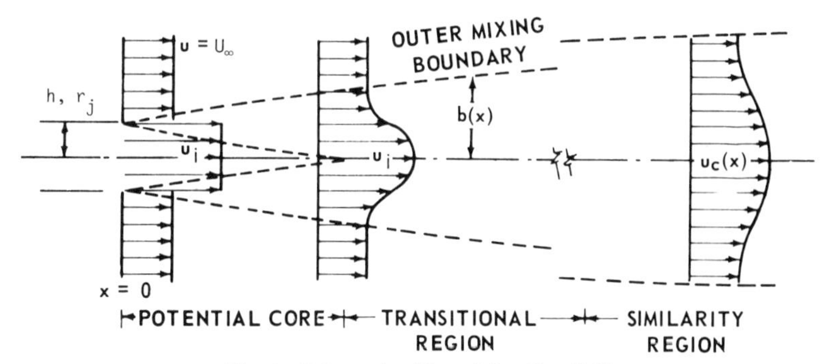

Fig. 1 Schematic of jet mixing flowfields.

shear layers from the edges of the jet merge on the axis. This is often called the potential core, since the flow of the jet fluid in this region is inviscid. Of course, if the flow coming out of the injector is fully developed, or nearly so, there is little or no potential core. Harsha[12] has given the following data correlation for air-air, axisymmetric jets:

$$x_c/d = 2.13\, Re_d^{0.097} \tag{31}$$

which can be used for estimating purposes. The data on the centerline velocity decay, $U_c(x/d)$, to be given in later sections will also show how the length of the initial region varies under different conditions.

2. *Mean-Flow Data in the Main Mixing Region for Constant-Density Flows*

Many workers have published data of this type for both the axisymmetric and planar geometries. For the axisymmetric geometry, the data of Antonia and Bilger[27] have been chosen as the major source for inclusion here for two reasons. First, some of the best known earlier works seem to have suffered from wall interference,[21] and, second, Ref. 27 also contains rather complete turbulence data which can be presented later to make a coherent package. The main parameter that characterizes these problems in a gross way is $m \equiv U_\infty / U_j$, where U_∞ is the external stream velocity and U_j is the jet velocity at injection. For severely distorted initial injection velocity profiles, U_j should be viewed as a mass-averaged value.

In Fig. 2, we show the variation of the centerline velocity in terms of $U_\infty / [U_c(x/d) - U_\infty]$ and a characteristic width $r_{1/2}$ defined by the relation

$$U(r_{1/2}, x) = [U_c(x) + U_\infty]/2 \tag{32}$$

with axial distance. For obvious reasons, this latter quantity is frequently called the half-radius, and it can be more easily determined accurately than

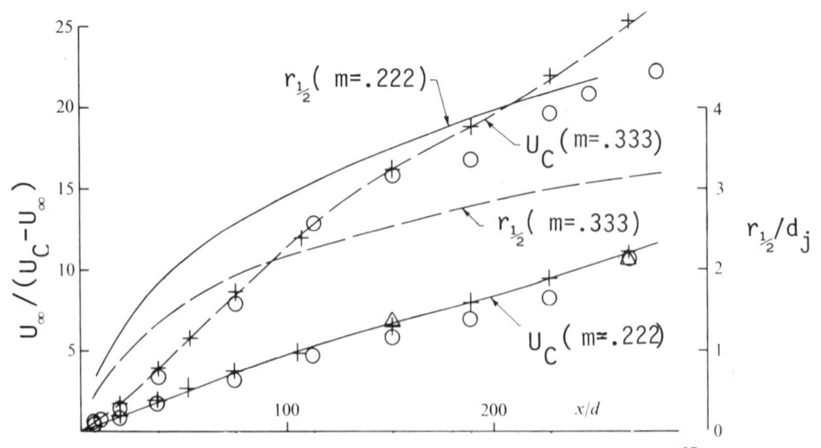

Fig. 2 Variation of centerline velocity and half-radius for an axisymmetric jet[27]: ○ △ hot wire; + pivot tube (published with permission).

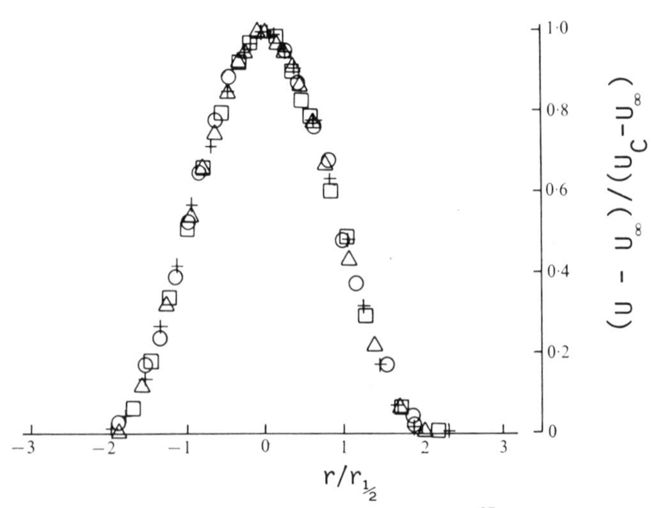

Fig. 3 Nondimensional radial profiles across an axisymmetric jet[27]: $m = 0.222$, $x/d_j = 38$ (+), 76 (△), 152 (□), 266 (○) (published with permission).

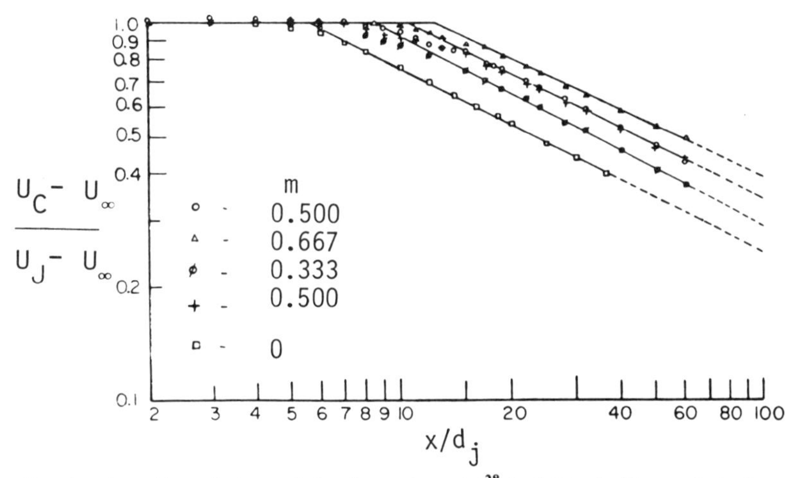

Fig. 4 Centerline velocity variation for a planar jet[28] (published with permission).

some ill-defined total width where the mixing zone merges asymptotically into the external stream. Several observations on these data are in order. First, the influence of the parameter m on the flowfield is quite profound. Second, the total distance to final decay where $U_c \to U_\infty$ is very long when measured in terms of jet diameters. Third, the agreement in the results obtained with a hot-wire anemometer and a simple pitot tube is quite good. Nondimensional, radial velocity profiles are given in Fig. 3 for various x/d. These data indicate that the profiles are apparently "similar" for $x/d \geq 40$. The term "similar" means that the profiles, expressed in terms of coordinates such as those on Fig.

3, remain unchanged with x/d. A word of caution here is, however, important. The coordinates used on Fig. 3 severely constrict the data, so that the satisfaction, or lack of it, of a similarity condition is difficult to assess. On such a plot, the value and the slope are fixed at $(r/r_{1/2})=0$, and the value is fixed at $(r/r_{1/2})=1$ and $(r/r_{1/2})\rightarrow\infty$. Considered in terms of classical analytic geometry, such curves are very strongly constrained.

For the planar geometry, the results of Weinstein et al.,[28] Bradbury,[29] and Everitt and Robins[30] will be utilized here. The variation of the centerline velocity with x for several values of m can be seen clearly on Fig. 4. Such data are often plotted on logarithmic scales, so that the potential core length (the region where the ordinate = 1.00) and the simple power law $(x/d)^P$ decay of the centerline velocity for large x/d can be displayed. Again, the strong influence of m on the flow is apparent. A similarity variable plot of transverse velocity profiles is shown in Fig. 5. These profiles also appear to be similar.

At this point, it is convenient to introduce a useful unifying concept. It is often assumed that far from the injector the precise details of the injector are unimportant, and all jets will behave in a like manner. The details of the injector can be accounted for in an "effective origin shift" x_0 that can be determined empirically or by scaling x with the momentum thickness of the flow at $x=0$ rather than with h or r_j. The success of this last idea is shown in Figs. 6a and 6b, where a considerable unification of all of the results can be seen.

The last mean-flow variable of importance is the static pressure. It has been common to assume that the static pressure is uniform across the mixing layer and also in the axial direction when there is no imposed, external pressure

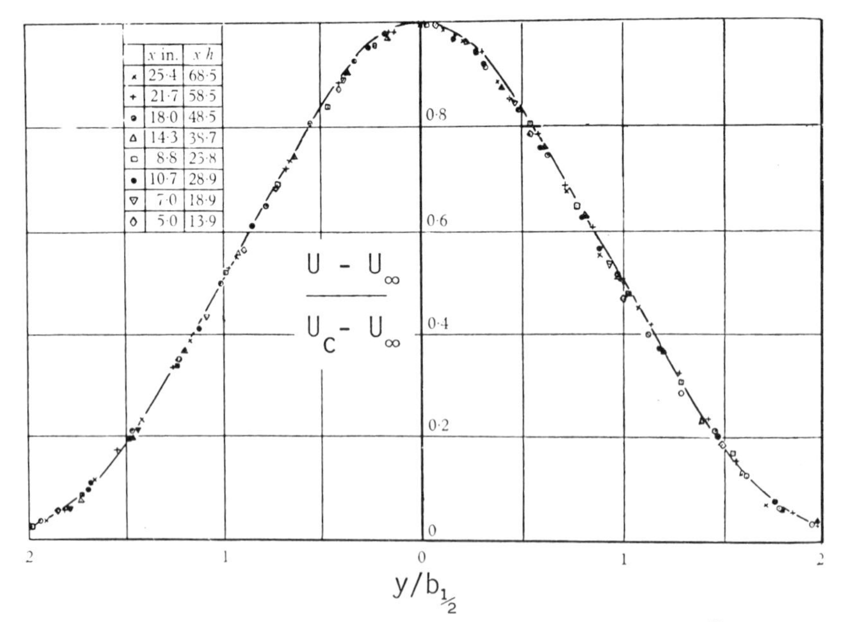

Fig. 5 Similarity plot of transverse profiles of axial velocity in a planar jet[29] with $m=0.16$ (published with permission).

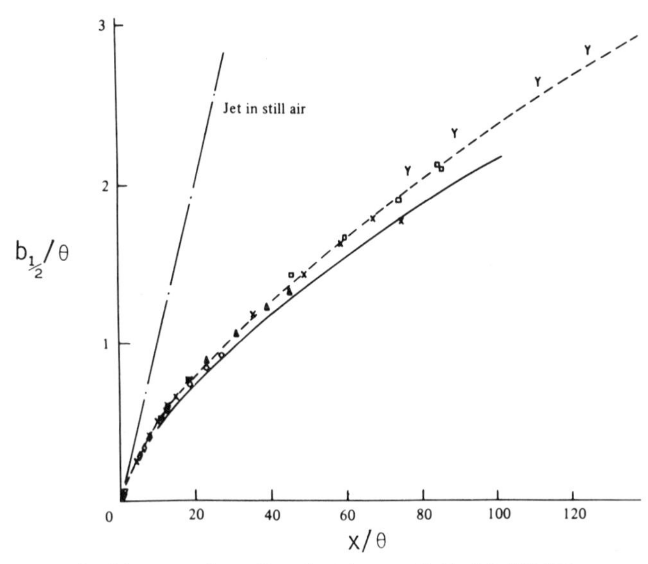

a) Streamwise Variation of Half-Width

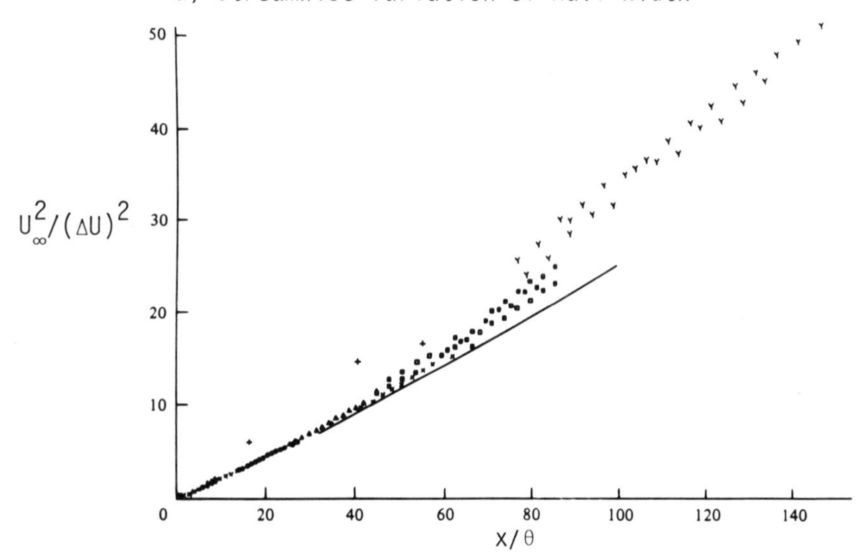

b) Streamwise Variation of Centerline Velocity

Fig. 6 Development of the flow from planar jets[30]: $U_j/U_\infty = 17.08$ (▽), 6.72 (◇), 4.48 (○), 3.78 (△), 3.29 (×), 3.24 (⊗), 3.03 (□), 2.60 (Y), 1.72 (+) from Ref. 29 (published with permission).

gradient. Experiment, however, indicates that this is not the case, as can be seen from the data on Fig. 7 which are for a planar jet with $U_\infty \ll U_j$.[29] The results of Ref. 31 for wake flows in various geometries do show that this behavior is much more pronounced in the planar than in the axisymmetric or three-dimensional geometries.

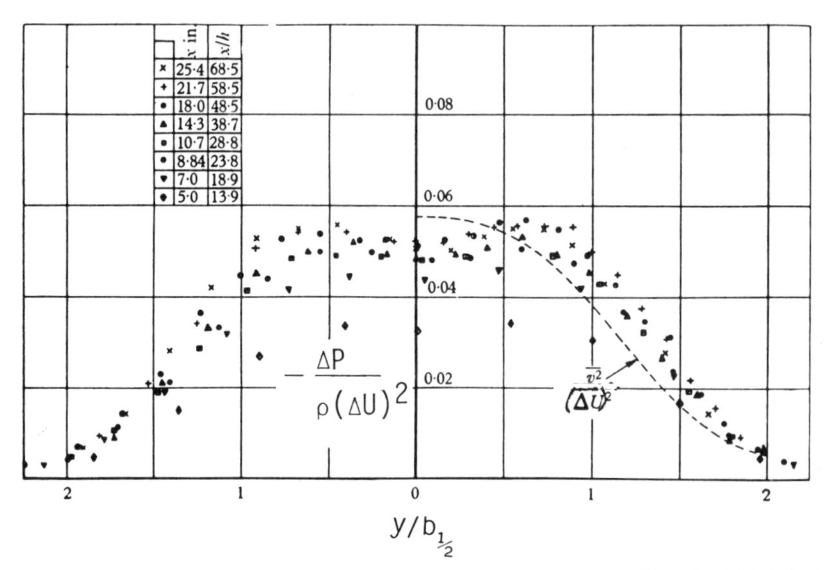

Fig. 7 Transverse profiles of static pressure in a planar jet with $m = 0.16$[29]: $x/h = 13.9$ (◆), 18.9 (▼), 23.8 (●), 28.8 (■), 38.7 (▲), 48.5 (⊙), 58.5 (+), 68.5 (×) (published with permission).

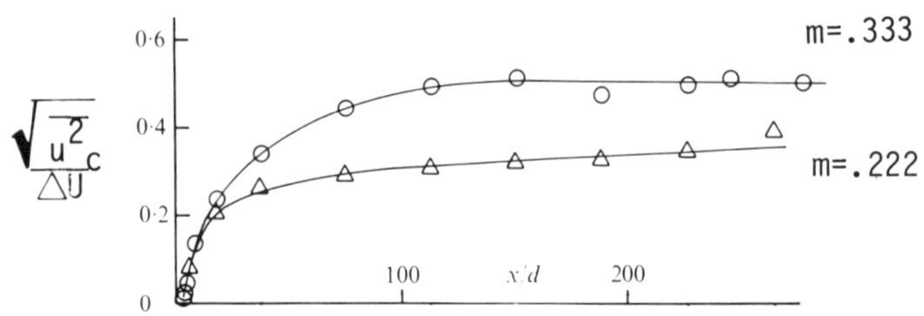

Fig. 8 Streamwise variation of centerline axial turbulence intensity in axisymmetric jets[27] (published with permission).

3. *Turbulence Data for Constant-Density Flows*

Many interested readers may have come to this point with the notion of turbulence as a very small, almost passive, fluctuating motion superimposed upon the main mean flow. Beginning students commonly adopt that erroneous concept. Perhaps that idea can be dispelled quickly by the information from Ref. 27 in Fig. 8. Here, we have the axis rms axial turbulence velocity nondimensionalized with the maximum mean-flow velocity difference across the mixing zone. The main message of interest to be gained from these data is that the average velocity fluctuations are of the same order as the mean-flow velocity difference. The situation is similar for other related flows, i.e., wakes behind a variety of bodies and jets into still air, as may be seen in the data collected in Ref. 32.

We can proceed now to an orderly presentation of some of the better turbulence data currently available. For the axisymmetric case, the data of

Antonia and Bilger[27] for the axial turbulence intensity are given in Figs. 8 and 9. The streamwise behavior of the axis values and the characteristic shape of the transverse profiles can be easily seen. The maximum generally occurs somewhat off the centerline. Note that the profiles of this variable have certainly not attained a similarity condition by $x/d \approx 200$. The data of Gibson[33] for tests with $U_\infty = 0$ are presented in Fig. 10 to show the relative magnitudes of the axial, transverse, and peripheral turbulence intensities. The variation of the turbulent kinetic energy across the layer, again for $U_\infty = 0$, is shown in Fig. 11 from Ref. 21 using the data of Refs. 34 and 35.

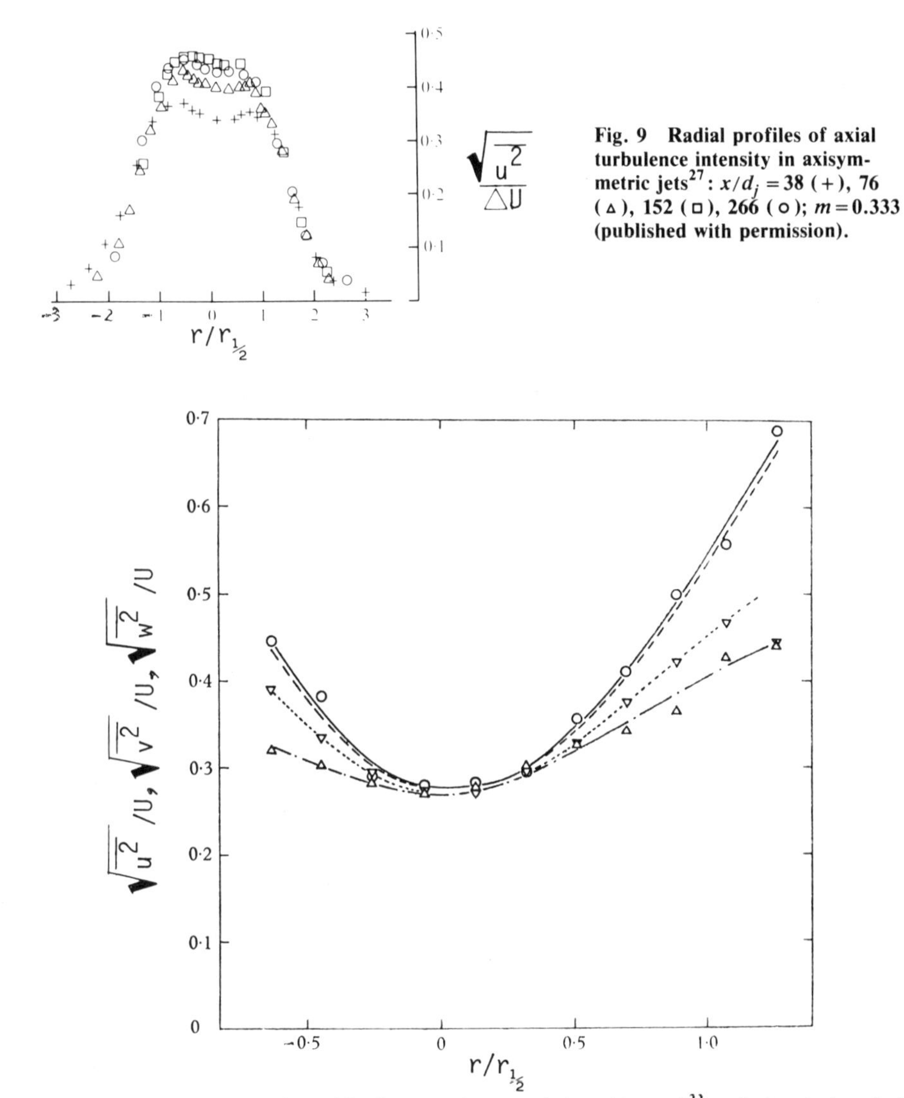

Fig. 9 Radial profiles of axial turbulence intensity in axisymmetric jets[27]: $x/d_j = 38$ (+), 76 (△), 152 (□), 266 (○); $m = 0.333$ (published with permission).

Fig. 10 Three turbulence intensities for an axisymmetric jet with $m = 0^{33}$: u (○), v (△), w (▽) (published with permission).

Information on the Reynolds turbulent shear stress, $\tau_T = -\rho\overline{uv}$, is available for $U_\infty \neq 0$ and is shown in Fig. 12. Values for the shear can also be inferred from a detailed knowledge of the mean-flow velocity variations via Eq. (8) (or in this case its axisymmetric equivalent); however, much greater precision is obtained by recasting the operation in terms of various integrals of the profiles and derivatives thereof. The details are given in Ref. 27, and some results are presented here on the bottom of Fig. 13 compared with directly measured values. Experimental values for the eddy viscosity can be found by using Eq. (12) and the experimental mean-flow profiles for $\partial U/\partial y$ and the turbulent stress measurements. They are generally given in terms of an eddy viscosity Reynolds number:

$$Re_T = \frac{(\Delta U) r_{1/2}}{\nu_T} \tag{33}$$

and some results are given here on the top of Fig. 13. These values of the shear stress are higher than those for $U_\infty = 0$,[35] lower than those reported for the

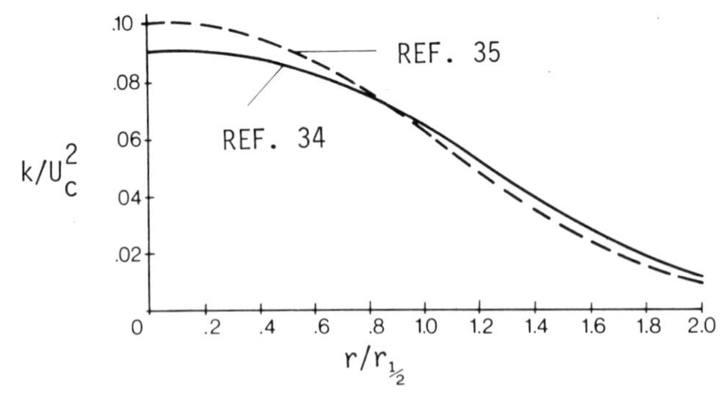

Fig. 11 Radial profile of turbulent kinetic energy for an axisymmetric jet with $m = 0^{21}$ (published with permission).

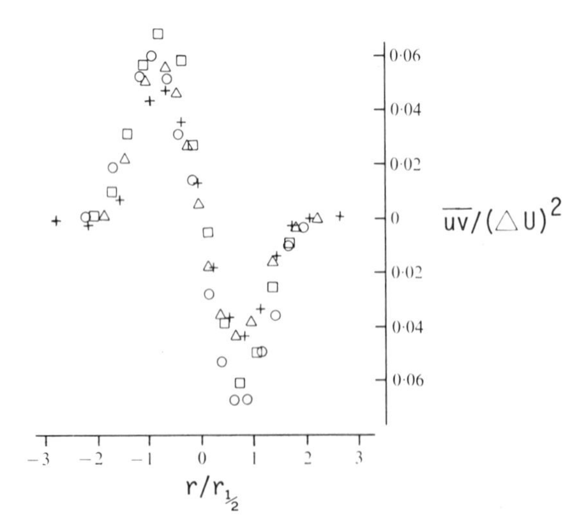

Fig. 12 Radial profiles of Reynolds stress for an axisymmetric jet with $m = 0.333^{27}$: $x/d_j = 38$ (+), 96 (△), 152 (□), 248 (○) (published with permission).

wake behind bluff bodies,[36] but of the same order as in the wake behind a streamlined body.[37] The converse of these statements is true concerning Re_T.

There are several other quantities that can be used to describe the turbulence which are of direct interest. Many of the quantities that follow were first studied experimentally and theoretically in isotropic and/or homogenous turbulence. Indeed, it is very difficult to obtain experimentally some of these values in actual shear flows without using the crucial interrelationships that rigorously apply only under those restrictions. We introduce first the velocity correlation coefficients:

$$f(r) \equiv [\overline{u_r(x) \cdot u_r(x+r)}] / (u')^2 \tag{34}$$

$$g(r) \equiv [\overline{u_n(x) \cdot u_n(x+r)}] / (u')^2 \tag{35}$$

Here r is the spacing between the sampling points, and u_r and u_n are the fluctuating velocities in the r direction and normal to it. For simplicity, consider a case where both sampling points are on the y axis. Because of the assumed local homogeniety of the flow, we expect, for example, $g(y) = g(-y)$. In fact, for small values of y, $g(y)$ is roughly parabolic. A

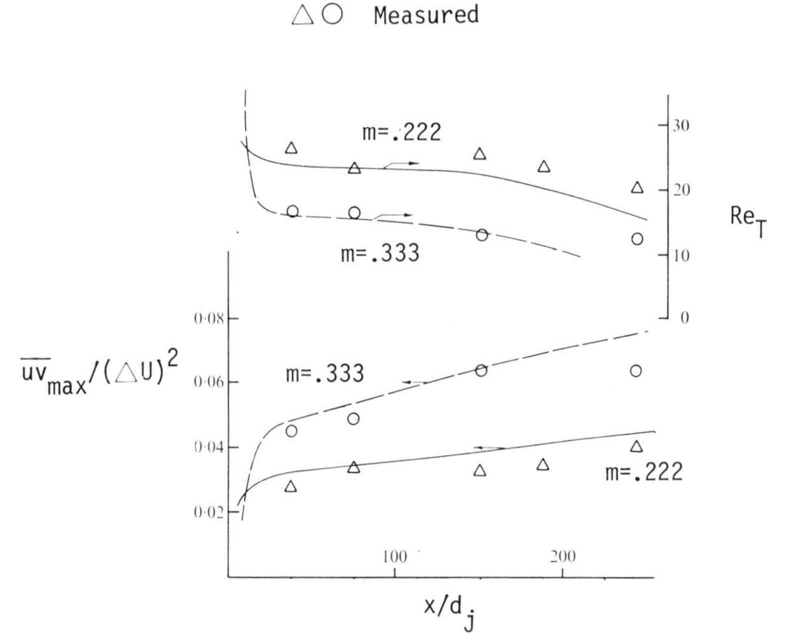

Fig. 13 Streamwise variation of maximum shear stress and eddy viscosity Reynolds number for an axisymmetric jet[27] (published with permission).

length scale λ_g can be introduced such that, for small y,

$$g(y) \approx 1 - (y^2/\lambda_g^2) \tag{36}$$

Expanding in a Taylor's series about $y = 0$, there results

$$\frac{1}{\lambda_g^2} = -\frac{1}{2}\left(\frac{\partial^2 g}{\partial y^2}\right)\bigg|_{y=0} = \frac{1}{2(u')^2}\left(\overline{\frac{\partial u}{\partial y}}\right)^2\bigg|_{y=0} \tag{37}$$

This length is called the microscale or the dissipation scale; it is a measure of the dimension of eddies that are mainly responsible for viscous dissipation. A second important length scale has been chosen to reflect the longest connection or the correlation distance between the velocities at two points. Clearly, the degree of correlation must decrease with increasing distance between the two points in question. The so-called integral scale, defined as

$$\Lambda_g = \int_0^\infty g(y)\,\mathrm{d}y \tag{38}$$

is generally taken as indicative of this scale of the turbulence. Analogous scales λ_f and Λ_f are also defined using Eq. (34) and definitions similar to Eqs. (37) and (38).

The length scales introduced earlier can also be defined through the use of the three-dimensional energy spectrum function:

$$\int_0^\infty E(K)\,\mathrm{d}K \equiv \tfrac{1}{2}(\overline{u^2} + \overline{v^2} + \overline{w^2}) \tag{39}$$

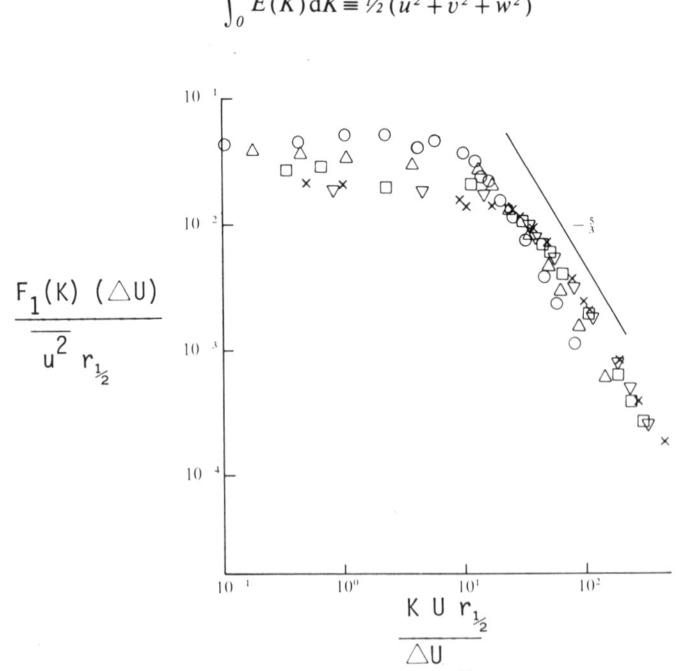

Fig. 14 Axial-component spectra for an axisymmetric jet[27]: x (cm) = 40 (○), 60 (△), 100 (□), 120 (▽), 140 (×); $m = 0.333$ (published with permission).

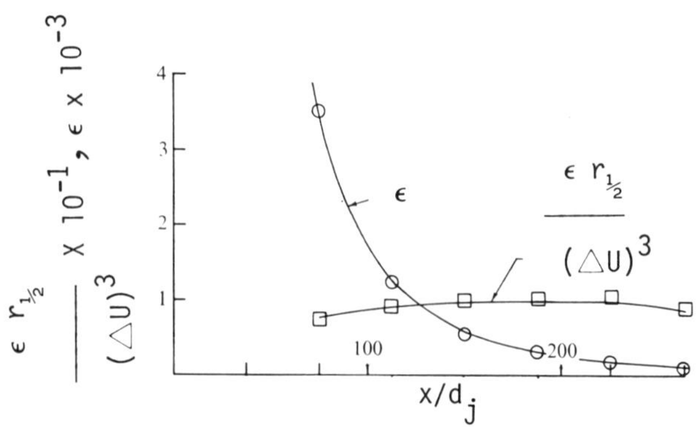

Fig. 15 Variation of dissipation rate along the centerline of an axisymmetric jet[27]: $m = 0.333$ (published with permission).

where $E(K)\,\mathrm{d}K$ is the contribution to the energy in the range of wave numbers between K and $K + \mathrm{d}K$. The one-dimensional energy spectrum functions are defined as

$$\int_0^\infty F_1(K)\,\mathrm{d}K = \overline{u^2} \tag{40a}$$

$$\int_0^\infty F_2(K)\,\mathrm{d}K = \overline{v^2} \tag{40b}$$

$$\int_0^\infty F_3(K)\,\mathrm{d}K = \overline{w^2} \tag{40c}$$

Assuming local isotropy, we write

$$\Lambda_f = \frac{\pi}{2}\,\frac{F_1(0)}{\overline{u^2}} \tag{41}$$

$$\lambda_f^2 = \frac{\tfrac{1}{3}\left(\overline{u^2} + \overline{v^2} + \overline{w^2}\right)}{\int_0^\infty K^2 F_1(K)\,\mathrm{d}K} \tag{42}$$

Finally, the rate of turbulent energy dissipation can be determined from

$$\epsilon = 2\nu \int_0^\infty K^2 E(K)\,\mathrm{d}K = 15\nu \int_0^\infty K^2 F_1(K)\,\mathrm{d}K \tag{43}$$

Various other methods have been discussed for determining ϵ (see Ref. 38), all of which also involve either severely restrictive assumptions, very indirect methods, or extreme experimental difficulty. Therefore, reported measurements of ϵ must always be accompanied by a statement as to how the data were obtained. Lastly, it is prudent to restate here that many of the relationships just given rigorously hold only under the assumptions of local isotropy and/or homogeniety. A thorough discussion can be found in Ref. 4.

As we proceed now to lay out the available information, it will be necessary to hop about with unfortunate frequency, since all of the desired data for each flow do not usually exist in any one reference. From the work of Ref. 27, data on the normalized, one-dimensional, axial energy spectrum are given in Fig. 14. The well-known (– 5/3) decay law predicted for the inertial subrange at the lower end of the universal range which is presumed to exist at wave numbers high enough for the small-scale turbulence to be statistically independent of the large-scale motion also is shown for comparison with the behavior of the data. The dissipation rate ϵ in Ref. 27 was estimated assuming that dissipation takes place almost completely within the equilibrium range, which leads to the relation

$$F_l(K) = A\epsilon^{2/3}K^{-5/3} \tag{44}$$

where A is approximately 0.5. The results are shown here in Fig. 15. A second dissipation length scale L_ϵ, defined in Ref. 27 as

$$L_\epsilon = (\overline{u^2})^{3/2}/\epsilon \tag{45}$$

also is shown on this graph. This and the other length scales Λ_f and λ_f are shown in Fig. 16.

With all of the types of mean-flow and turbulence information collected and presented in the preceding, a very important combined presentation can now be made. This is the energy balance across the flow where the magnitude of the individual terms in Eq. (17) can be seen and studied. The most complete and reliable information available is the basic data of Ref. 35 for $U_\infty = 0$ as modified by Rodi in Ref. 21. This balance is presented in Fig. 17. In this case, the energy production by the normal stresses

$$-(\overline{u^2} - \overline{v^2})\frac{\partial U}{\partial x} \tag{46}$$

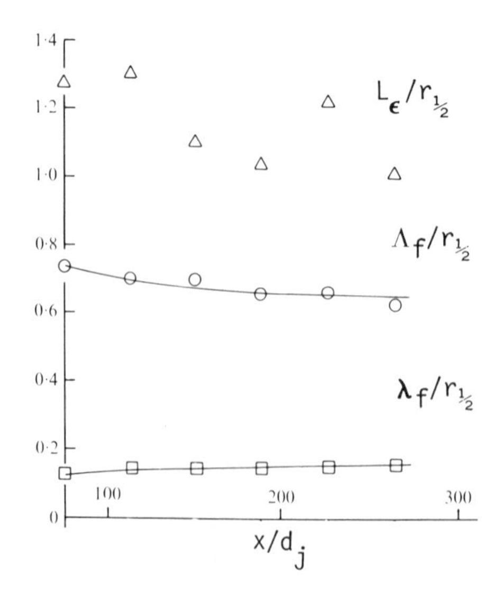

Fig. 16 Variation of turbulent length scales for an axisymmetric jet with $m = 0.333$[27] (published with permission).

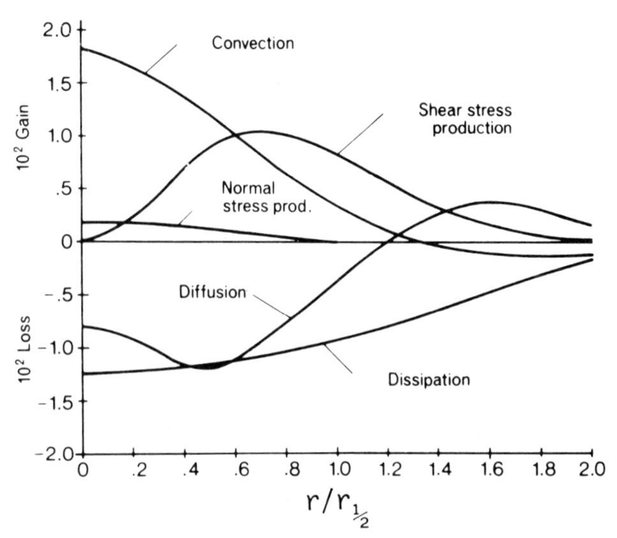

Fig. 17 Turbulent energy balance for an axisymmetric jet with $m = 0$ using modified data of Ref. 35.[21] All terms $\times\ r_{1/2}/U_c^3$ (published with permission).

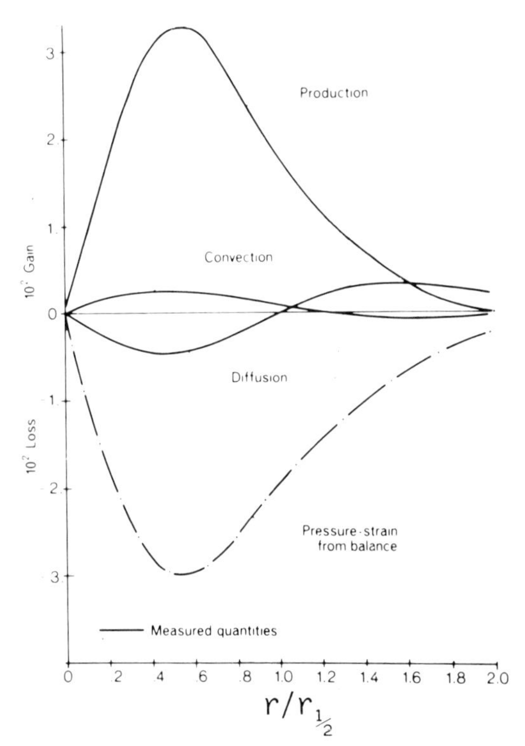

Fig. 18 Shear stress balance for an axisymmetric jet with $m = 0$ constructed with the data of Ref. 35.[21] All terms $\times\ r_{1/2}/U_c^3$ (published with permission).

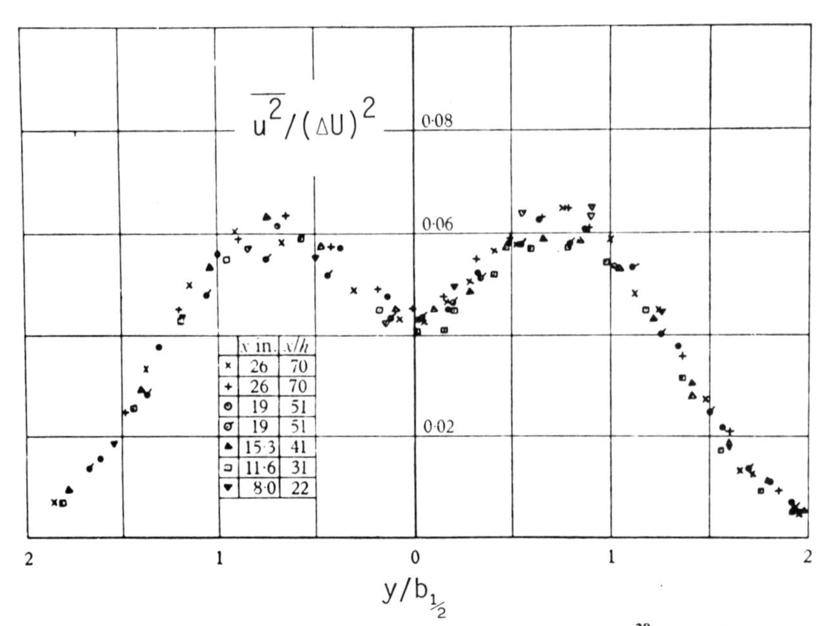

Fig. 19 Transverse profiles of axial turbulence intensity in a planar jet[29] (published with permission).

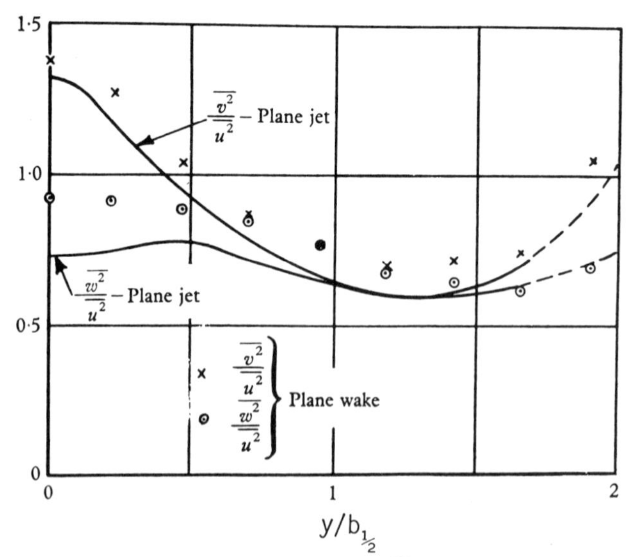

Fig. 20 Ratios of turbulent intensities in a planar jet.[29] Planar wake values are from Ref. 24 (published with permission).

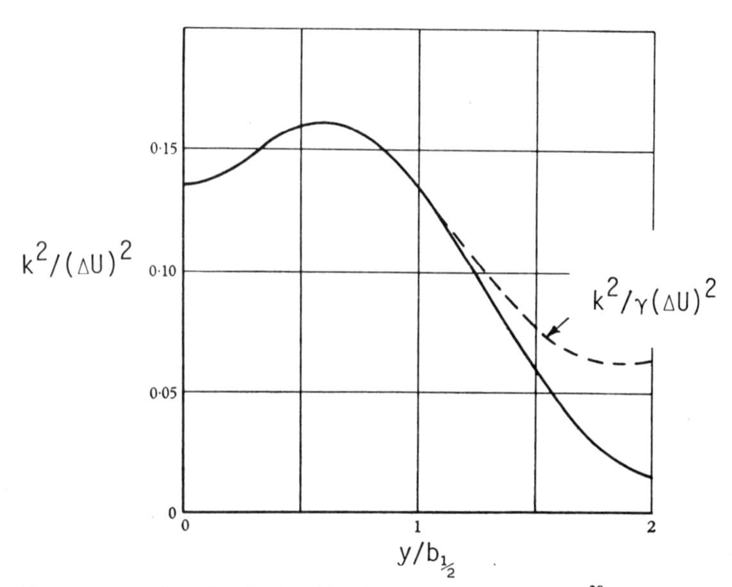

Fig. 21 Transverse profile of turbulent kinetic energy in a planar jet[29] (published with permission).

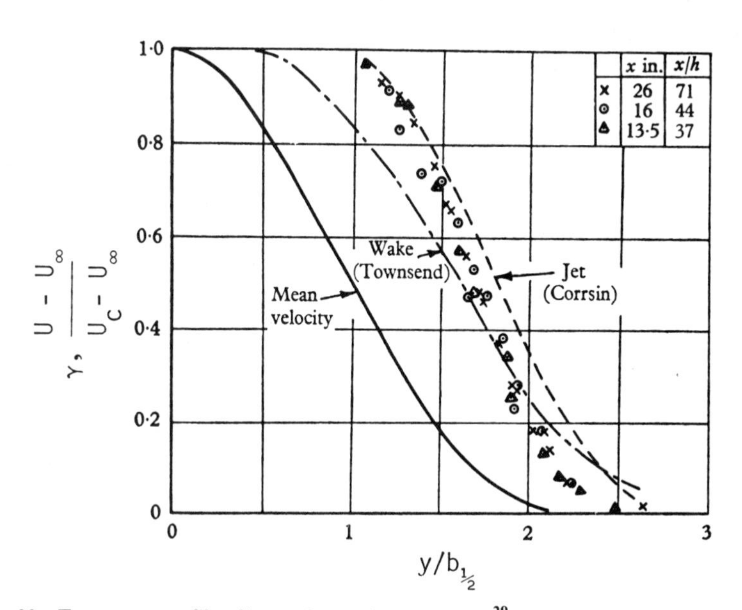

Fig. 22 Transverse profile of intermittency in a planar jet[29] (published with permission).

was also included for completeness, in addition to production by shear stress. It can be seen that this contribution is the smaller of the two, except near the axis. Finally, a shear stress balance can also be constructed. The information on Fig. 18 is again taken from Ref. 21.

We turn now to planar cases and look at the same types of turbulence data as were presented for the axisymmetric geometry. Profiles of the axial turbulence intensity from Ref. 29 are given in Fig. 19 on a similar basis as the axisymmetric data in Fig. 9, except that the ordinate here is the square of that in Fig. 9. The centerline values are perhaps slightly lower in the planar case for the same m. The ratio of the other turbulence intensities to the axial are shown in Fig. 20. The distribution of the turbulence kinetic energy across the layer for the same planar case is plotted on Fig. 21, as well as that quantity divided by the "intermittency" γ. This is defined as the fraction of time that the flow is locally turbulent. As the outer edge of any shear layer is approached, an irregular, highly unsteady boundary is encountered, and an observer is alternately in a turbulent eddy or in the freestream. Thus the intermittency approaches zero. The measured distribution for γ is shown on Fig. 22 with those for a planar wake and an axisymmetric jet with $U_\infty = 0$.

The measured Reynolds stress profiles and the values inferred from the mean-flow profiles are given in Fig. 23. These values are also rather lower than the corresponding ones in the axisymmetric case. The variation of the eddy viscosity across the layer in terms of the eddy Reynolds number is displayed in Fig. 24, and again these nondimensional eddy viscosities are lower than for the round jet. A one-dimensional spectrum in terms of frequency is plotted in Fig. 25. Again the $(-5/3)$ law serves well for the smaller eddies.

In Ref. 29, the dissipation rate was determined by measuring the mean square of the time rate of change of u and using

$$\epsilon = 15\nu\overline{(\partial u/\partial t)^2} \tag{47}$$

This procedure requires a local isotropy assumption. The results are shown in Fig. 26, along with the other parts of the turbulent energy balance. The behavior of the various terms in the balance is quite different in the planar and axisymmetric cases, as can be seen by comparing Figs. 17 and 26. The values of the integral length scales for various cases including the planar jet are shown in Table 3.

No complete, reliable shear stress balance is available from any one study, and so Rodi[21] has constructed the composite shown in Fig. 27. In this case, the

Table 3 Values of integral length scales

Flow	Author	Λ_g/b	Λ_f/b	Λ_f/Λ_g
Isotropic turbulence	Theoretical result	...	...	2.0
Plane jet	Bradbury (1965)[29]	0.38	...	...
Plane wake	Townsend (1956)[24]	0.6	0.83	1.37
Circular jet	Corrsin (1943)[39]	0.23	...	...
Circular jet	Corrsin and Uberoi (1949)[40]	0.17	0.92	5.4
Circular jet	Gibson (1963)[33]	...	0.95	...

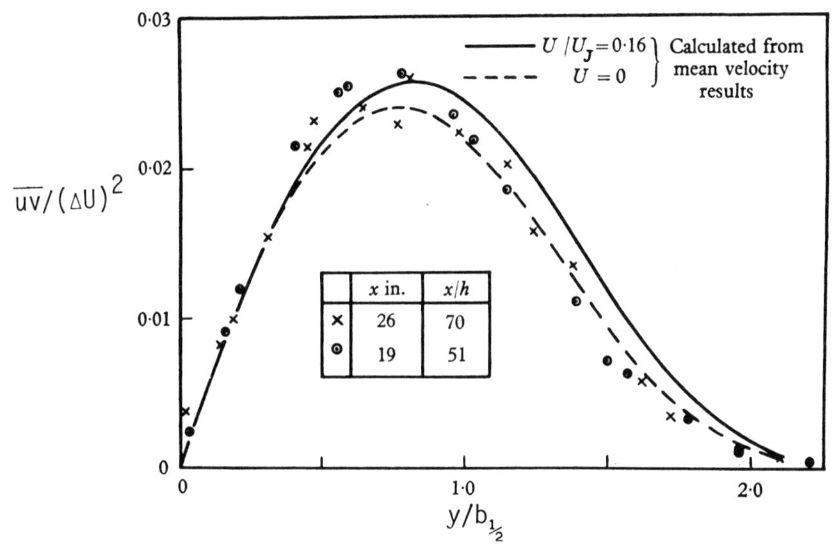

Fig. 23 Transverse profile of Reynolds stress in a planar jet[29] (published with permission).

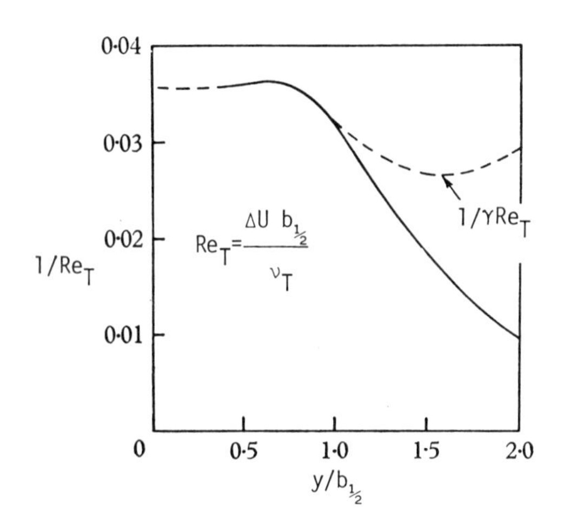

Fig. 24 Transverse profile of eddy viscosity Reynolds number in a planar jet[29] (published with permission).

planar and axisymmetric results are quite similar. The variation of the microscale across the layer is plotted in Fig. 28.

Some of the results of Ref. 30 do not agree with those in Ref. 29. Part of the differences can be attributed to initial conditions and part to measurement difficulties. However, it is suggested in Ref. 30 that the differences may be real and important, and further studies with other instrumentation are recommended.

4. *Effects of Temperature Variations*

The simplest cases with temperature variations are produced with injection at a temperature different from that of the main flow. Experiments of that

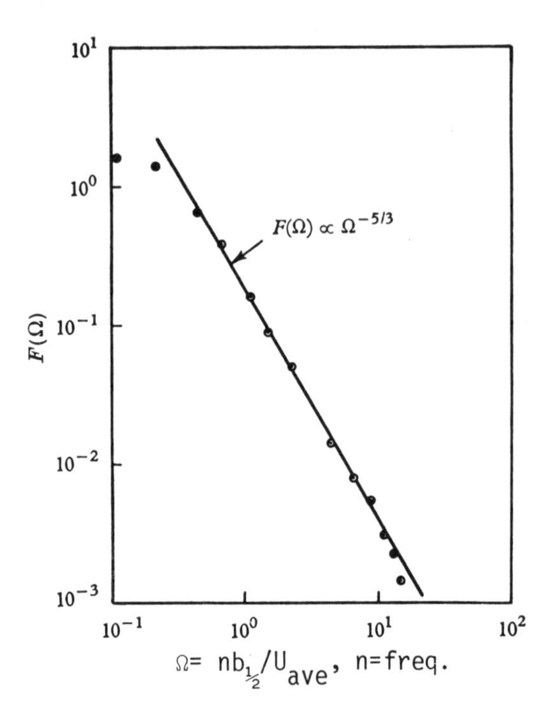

Fig. 25 $\overline{u^2}$ **spectrum in a planar jet showing an inertial subrange[29] (published with permission).**

type in the axisymmetric geometry are reported in Ref. 41. Mean-flow results in terms of the axial variation of the centerline velocity and temperature are shown here in Fig. 29 for a case with $T_\infty/T_j = 0.77$ and $U_\infty/U_j \equiv m = 0.50$. (The density ratio $\rho_\infty/\rho_j \equiv n$ for such cases is related to the temperature ratio as $T_\infty/T_j = \rho_j/\rho_\infty = 1/n$.) It can be seen that the power decay law exponent P is slightly greater for this case than that for the nearly constant density case presented in Ref. 42. The results for all of the tests and those of some other workers are listed in Table 4, where it is seen that a clear pattern does not emerge. The entries with a "×" in the C column are not of interest here but will be referred to later in Sec. II.B.5.

The axial variations of the half-widths for the velocity and temperature profiles are shown in Fig. 30 for the same case as in Fig. 29. From these data, we can see the important result that nondimensionalized temperature profiles are always "fuller" than the corresponding velocity profiles, which indicates that the turbulent transport of thermal energy is more rapid than that for momentum. This means that a turbulent thermal conductivity κ_T defined by

$$-\kappa_T \frac{\partial T}{\partial y} = \rho C_p \overline{vT'} \tag{48}$$

when combined with ν_T to give a turbulent Prandtl number

$$Pr_T \equiv \frac{\nu_T}{\kappa_T/\rho C_p} \tag{49}$$

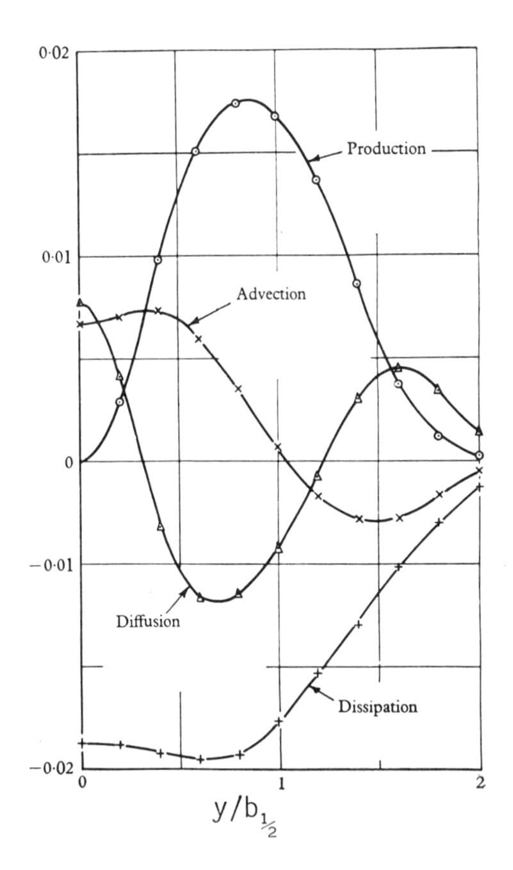

Fig. 26 Turbulent kinetic energy balance across a planar jet[29] (published with permission).

will have values $Pr_T < 1$. For the axisymmetric jet case in Ref. 40, this quantity was determined to be approximately 0.7. Actually, the value of the turbulent Prandtl number has been found to vary across the profile, especially near the outer edge, where intermittency becomes important.[44] Some results obtained from the measurements of Ref. 43 are given here in Fig. 31. Very interesting results of the same type are plotted in Fig. 32 for jets of mercury and oil from Ref. 45. These are significant, since the laminar Prandtl numbers for oils and mercury span the range $Pr \ll 1$ to $Pr \gg 1$.

Cases with a planar geometry have not been as widely studied with variable temperature. The investigation of Ref. 46 does provide velocity and temperature profiles and values of Pr_T for the special case $U_\infty = 0$. The profiles are displayed in Fig. 33, along with some for the axisymmetric case, also with $U_\infty = 0$. The planar data indicate a value of $Pr_T = 0.5$. This value is also confirmed by Ref. 44, and the difference in Pr_T for the axisymmetric and planar geometries has never been fully explained.

For turbulence data on heated jets, the recent work of Ref. 47 with $U_\infty = 0$ and an axisymmetric arrangement has been chosen. This study was done using a two-wire system to measure velocity and temperature fluctuations simultaneously but separately. When an X-wire arrangement was necessary for

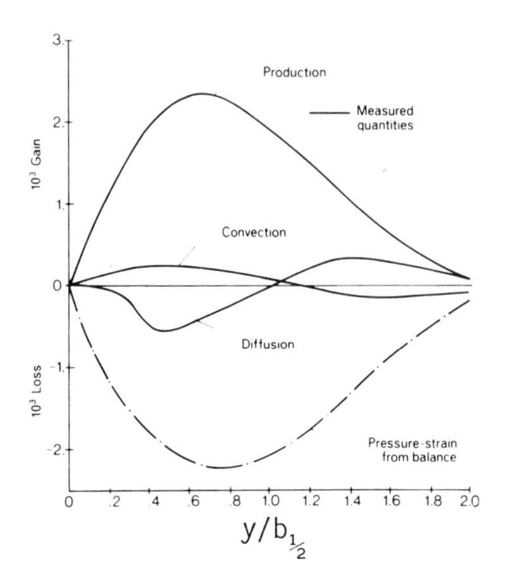

Fig. 27 Shear stress balance for a planar jet with $m = 0$ using the data of several workers.[21] All terms $\times\ b_{1/2}/U_c^3$ (published with permission).

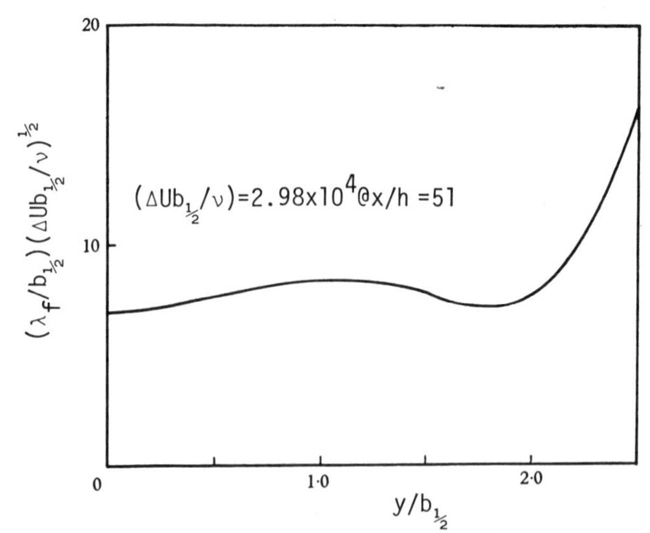

Fig. 28 Transverse profile of turbulent microscale in a planar jet[29] (published with permission).

the velocity field (e.g., to measure $\overline{uv}$), a third wire was used for temperature. Velocity and temperature fluctuations at $x/d = 15$ are given here on Fig. 34. The radial variations of turbulent shear and heat transfer are shown in Fig. 35. These profiles have similar shapes, and their peaks occur at roughly the same radial station, which is not, however, at the location of the maximum value of either $\partial U/\partial y$ or $\partial T/\partial y$. These facts suggest that the transport mechanisms for momentum and heat are quite similar and that gradient transport models are suspect even for this simple flow problem.

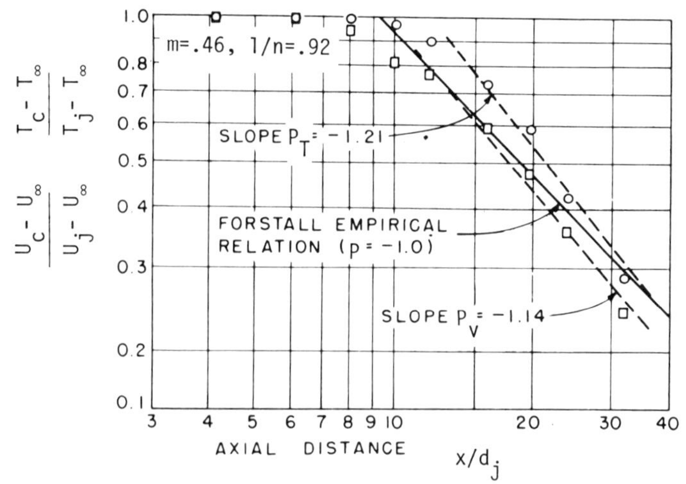

Fig. 29 Streamwise variation of centerline velocity and temperature in a heated, axisymmetric jet[41] (published with permission).

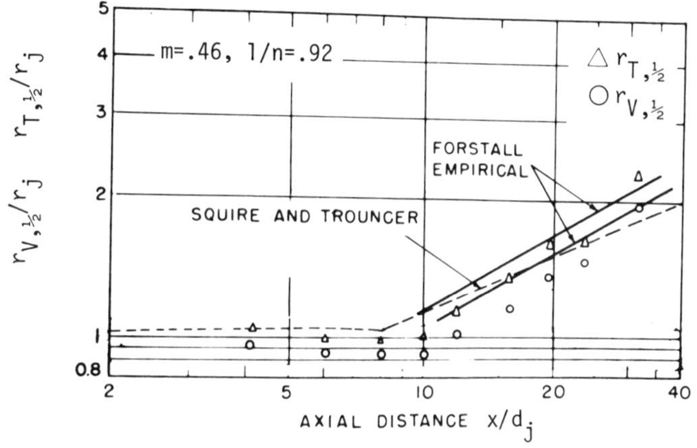

Fig. 30 Variation of half-radii based on velocity and temperature in a heated, axisymmetric jet[41] (published with permission).

Finally, high-speed flow situations produce substantial temperature variations. There has been no evidence to indicate, however, that significant new processes occur beyond those that occur in low-speed, heated jets, at least as far as the mean flow is concerned.

5. *Effects of Composition Variations*

Many important practical applications involve injection of one fluid into surroundings of a different fluid. In this section, we shall restrict ourselves to cases of one phase, either gas or liquid, and consider axisymmetric cases first.

An experimental correlation for the length of the potential core in terms of concentration of injected species was developed in Refs. 48 and 49. Both had

Table 4 Exponent P describing axial decrement

Investigator	m	$1/n$	V[a]	T[b]	C[c]	P_v	P_T	P_c
Corrsin[39]	0	0.95	×	×	...	−1.0	−1.0	...
	0	0.52	×	×	...	−1.18	−1.20	...
Forstall[42]	0.25	0.92	×	...	×	−1.0	...	−1.0
	0.50	0.92	×	...	×	−1.0	...	−1.0
Pabst[43,d]	0.05	0.45	×	×	...	−1.29	−1.05	...
	0.26	0.45	×	×	...	−1.28	−0.94	...
	0.44	0.45	×	×	...	−0.88	−0.74	...
Landis and Shapiro[41]	0.25	0.92	×	×	...	−1.00	−1.00	...
	0.25	0.77	×	×	...	−1.04	−1.04	...
	0.46	0.92	×	×	...	−1.21	−1.14	...
	0.50	0.77	×	×	...	−1.03	−1.08	...
	0.50	0.84	×	×	...	−1.33	−1.33	...
	0.50	0.84	×	...	×	−1.02	...	−1.02
	0.50	0.84	×	×	×	−1.22	−1.22	−1.22

[a] V = velocity (momentum) transfer.
[b] T = temperature variation.
[c] C = species variation.
[d] The results of Pabst correspond to a Mach number of 0.75.

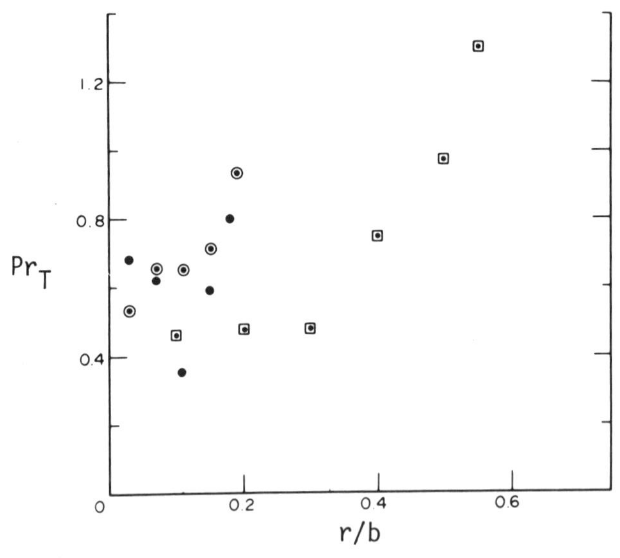

Fig. 31 Radial profile of turbulent Prandtl number for heated, axisymmetric jets with $m \neq 0$ using the data of Ref. 43[44] (published with permission).

the form

$$x_c/r_j = \text{const} \times -\sqrt{\rho_j U_j/\rho_\infty U_\infty} \tag{50}$$

but they disagreed on the value of the constant. Reference 48 had 22.0, and Ref. 49 gives a value of 13.0. A comprehensive study of the mixing of air, helium, and Freon-12 jets in an airstream was presented in Ref. 50. Some of

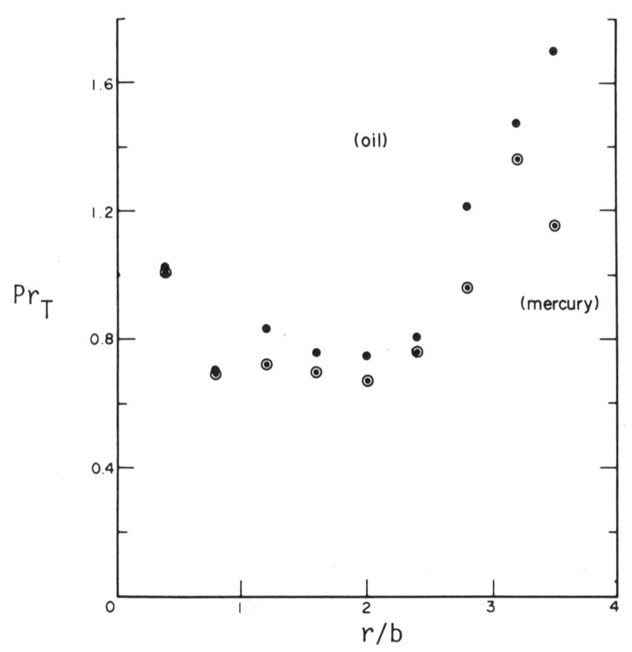

Fig. 32 Radial profiles of turbulent Prandtl number for heated, axisymmetric jets of oil and mercury with $m = 0$ using the data of Ref. 45[44] (published with permission).

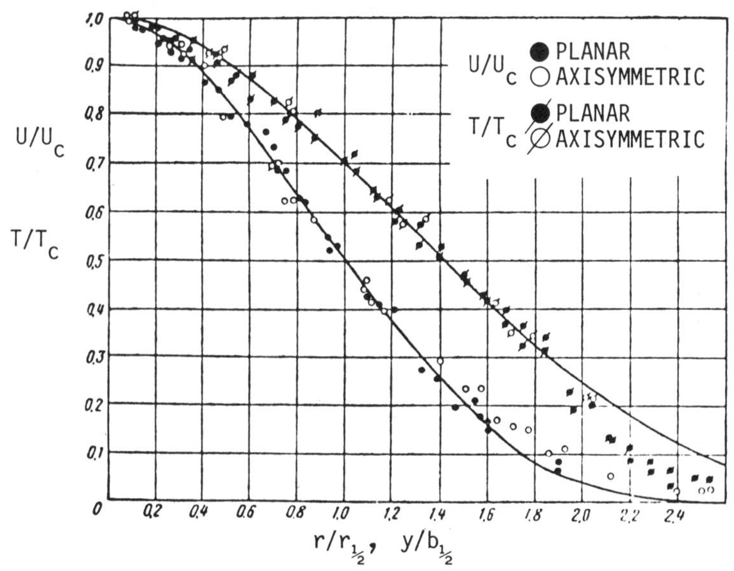

Fig. 33 Transverse velocity and temperature profiles for planar and axisymmetric jets with $m = 0$[46] (published with permission).

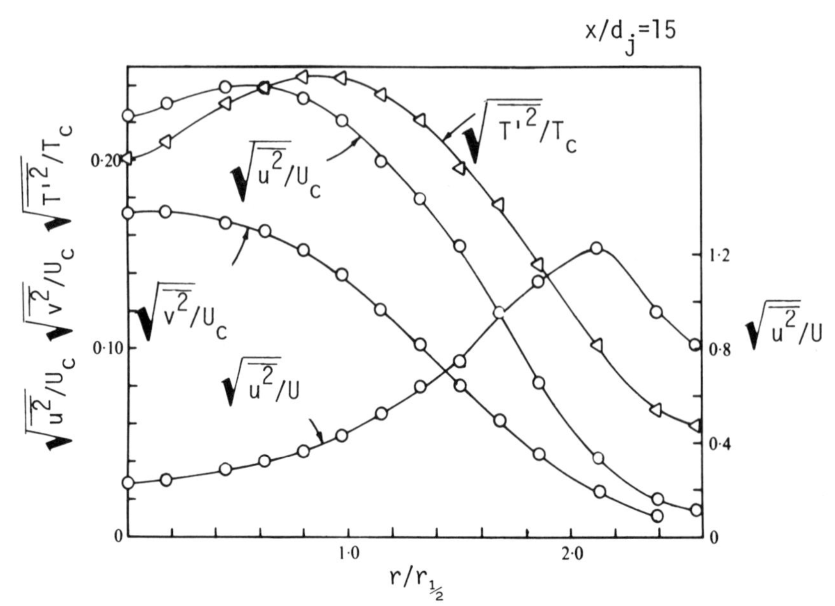

Fig. 34 Radial profiles of velocity and temperature fluctuations for an axisymmetric jet with $m = 0^{47}$ (published with permission).

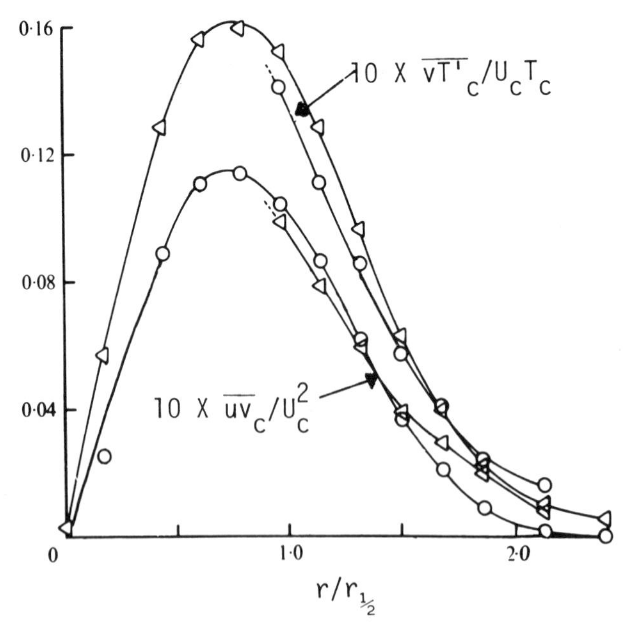

Fig. 35 Radial profiles of turbulent shear and heat flux for an axisymmetric jet with $m = 0^{47}$ (published with permission).

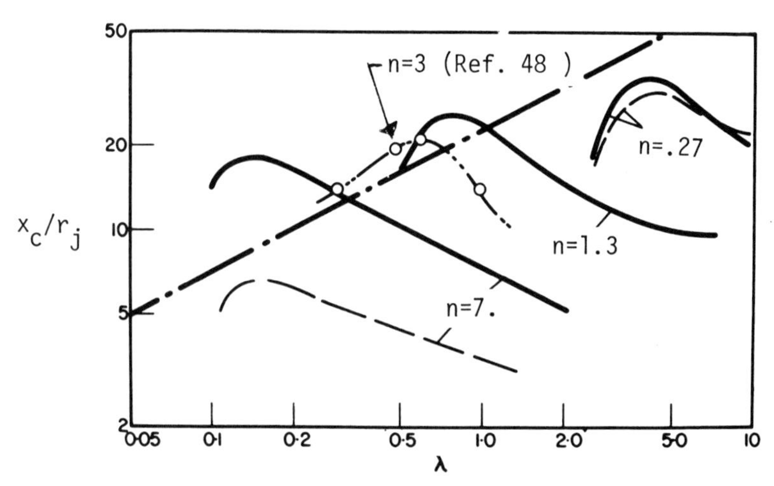

Fig. 36 Variation of the length of the initial region for heated and/or variable-composition axisymmetric jets[50] (published with permission).

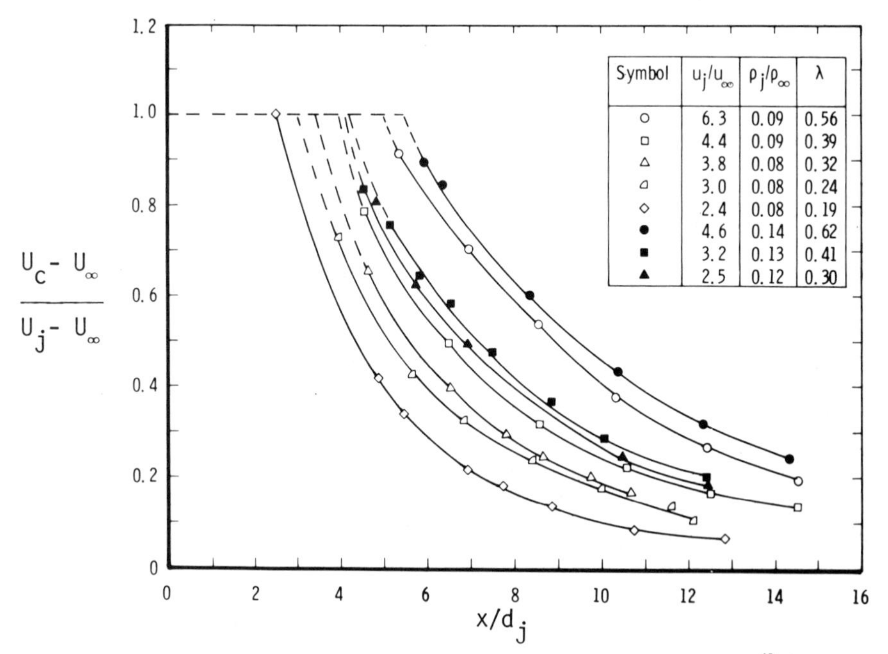

Symbol	u_j/u_∞	ρ_j/ρ_∞	λ
○	6.3	0.09	0.56
□	4.4	0.09	0.39
△	3.8	0.08	0.32
◸	3.0	0.08	0.24
◇	2.4	0.08	0.19
●	4.6	0.14	0.62
■	3.2	0.13	0.41
▲	2.5	0.12	0.30

Fig. 37 Streamwise variation of centerline velocity for axisymmetric H_2 jets into air[49] (published with permission).

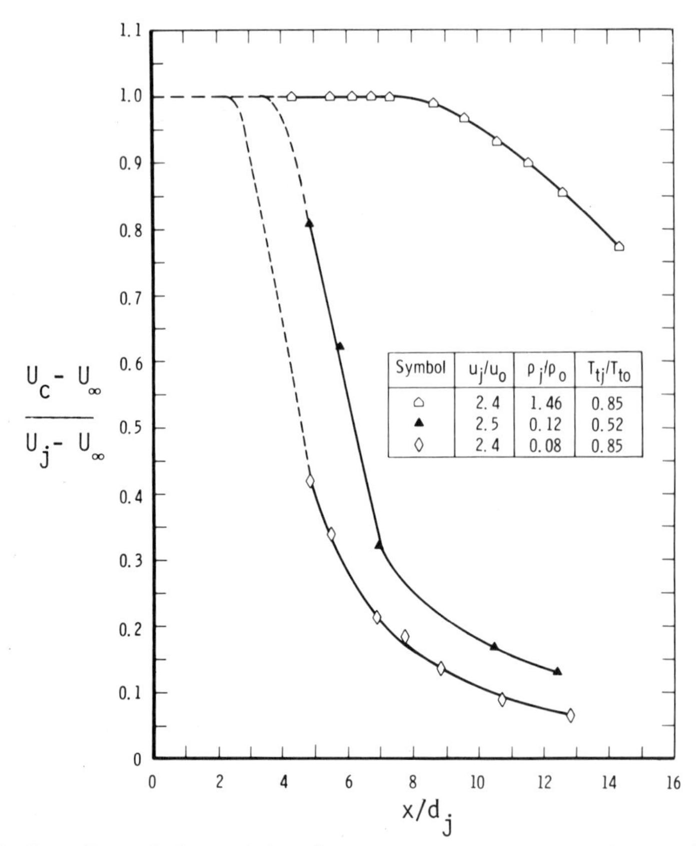

Fig. 38 Centerline velocity variation for a variable-composition axisymmetric jet with $m=$ const[49] (published with permission).

the injectants were also heated, so that the density ratio n covered the entire range from 0.27 to 8.2. One result of interest is a disagreement with the adequacy of Eq. (50). Abramovich et al.[50] present data to indicate that $m=U_\infty/U_j$ and $n=\rho_\infty/\rho_j$ are influential separately, as shown in Fig. 36.

For data on the influence of composition variations on the axial decay of centerline flow variables, we turn back to Ref. 49. In Fig. 37, we have the influence of n and m on the centerline velocity decay. The influence is obviously strong. This fact is clearly shown in Fig. 38, where the initial velocity ratio m is held constant, but n is varied over a wide range. The axial variation of the centerline composition of injectant is presented in Fig. 39.

When the axial decay of centerline values of the data is plotted on log paper, a power law decay rate is again shown. Many workers have attempted to determine the values of the relevant exponents for each type of variable: velocity, temperature, and composition. The results of Ref. 40 are listed in Table 4. The investigations of Ref. 50 indicated $0.85 \leq P_v \leq 1.25$ for $0.27 \leq n \leq 7.2$, $P_T \approx 1.3$, and $P_c \approx 1.0$, with the parameter m having little influence, all for the range studied. In Ref. 51, a large group of data from

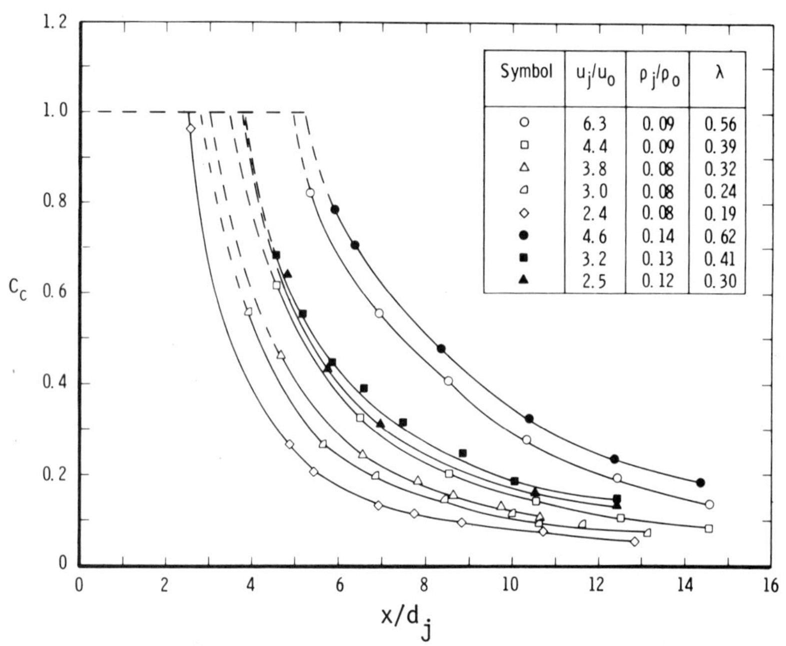

Fig. 39 Streamwise decay of centerline concentration for an axisymmetric H_2 jet into air[49] (published with permission).

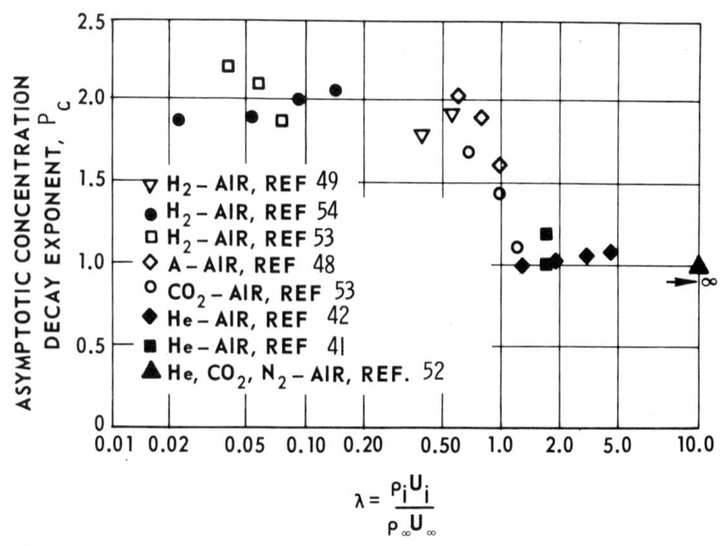

Fig. 40 Experimental determination of asymptotic concentration decay exponent.

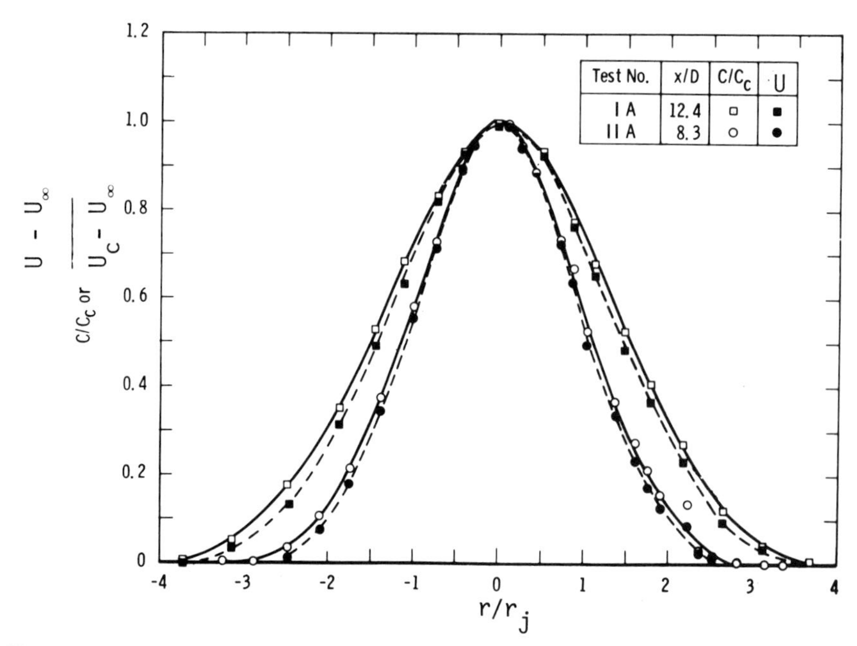

Fig. 41 Radial profiles of velocity and concentration for an axisymmetric H_2 jet into air[49] (published with permission).

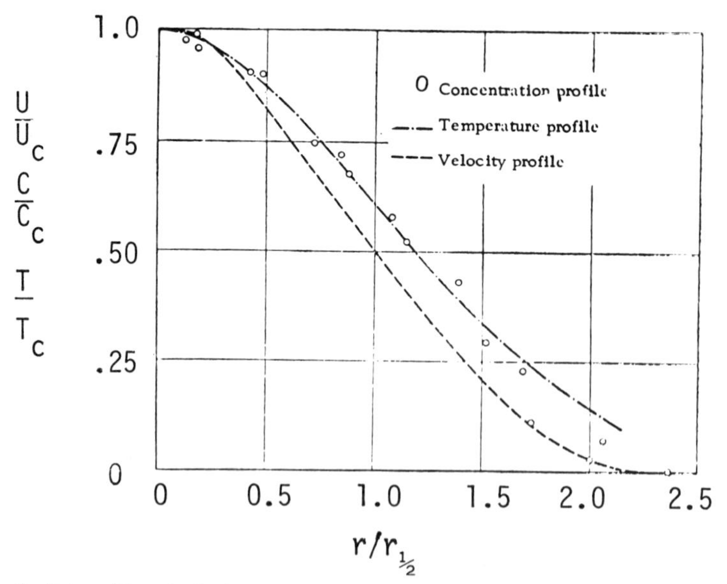

Fig. 42 Radial profiles of velocity, temperature, and concentration for an axisymmetric jet with various *m* from Refs. 42 and 55[5] (published with permission).

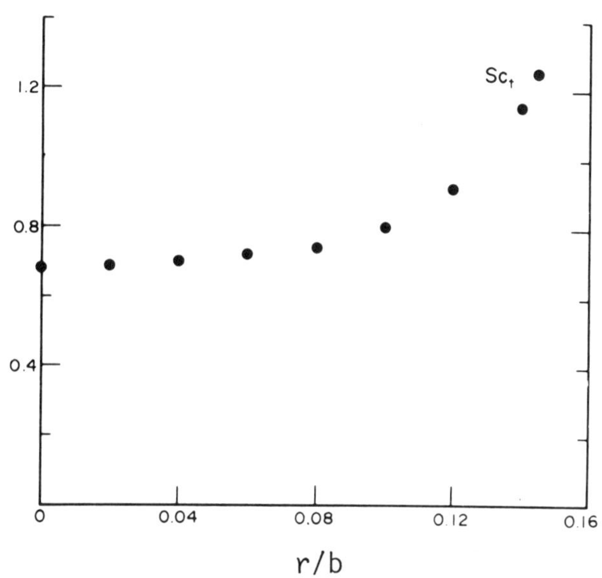

Fig. 43 Radial profile of turbulent Schmidt number for an axisymmetric jet with $m = 0$[44] (published with permission).

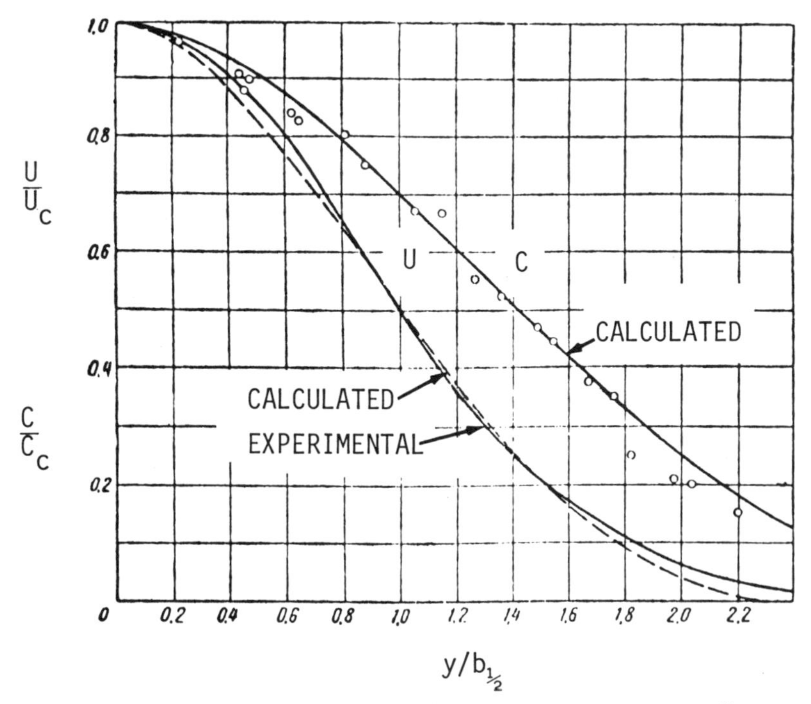

Fig. 44 Transverse profiles of velocity and concentration for a planar jet with $m = 0$[46] (published with permission).

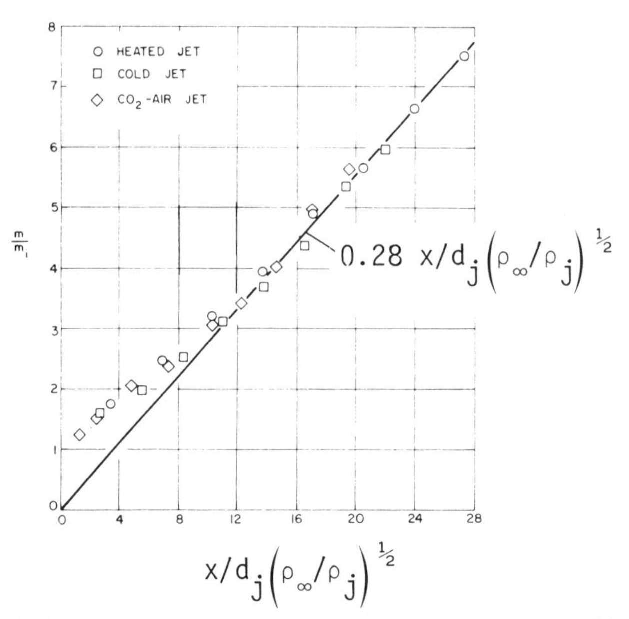

Fig. 45 Correlation of mass entrainment data for axisymmetric jets with $m = 0$[57] (published with permission).

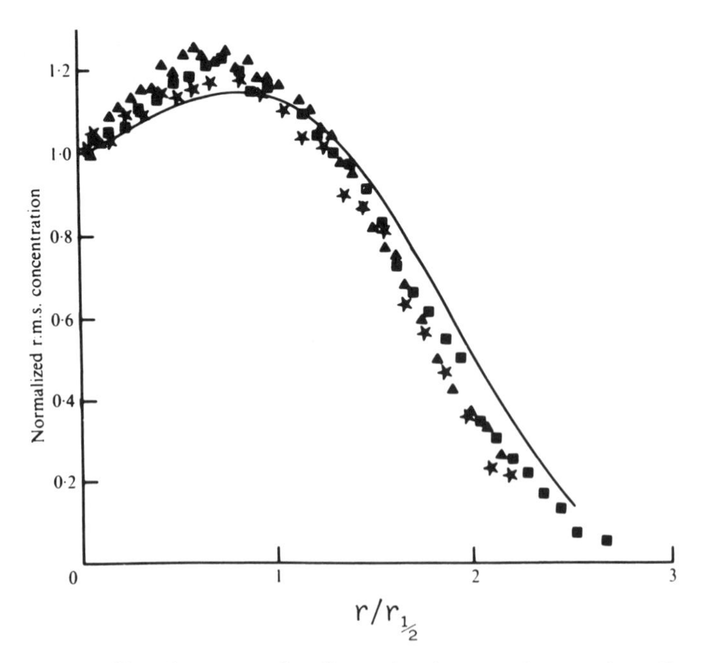

Fig. 46 Radial profiles of concentration fluctuation for an axisymmetric methane jet into stationary air[60]: $x/d_j = 20$ (■), 30 (▲), 40 (⁎);——Ref. 58 (published with permission).

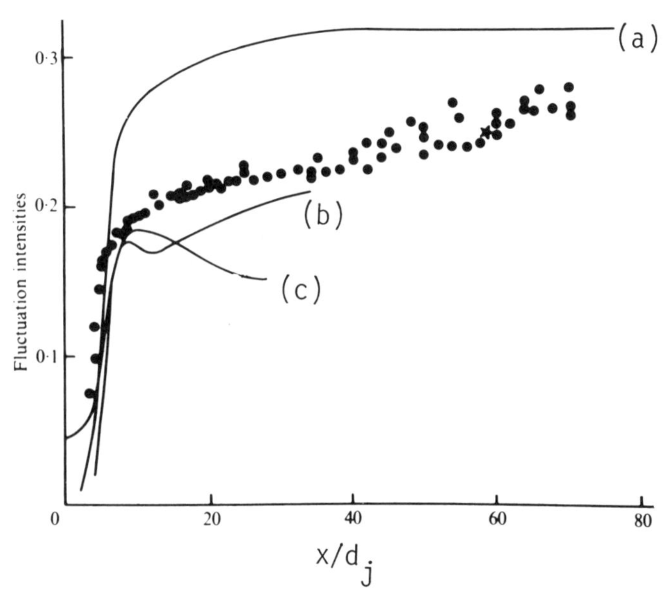

Fig. 47 Streamwise variation of centerline fluctuation intensities for an axisymmetric jet with $m=0$: concentration, Ref. 60 (●); temperature, Ref. 61 (*); velocity, Ref. 60 (a); concentration, Ref. 58 (b); temperature, Ref. 40 (c)[60] (published with permission).

various investigators was analyzed to develop a correlation for P_c in terms of the single parameter $\lambda \equiv 1/mn$. The result is shown here in Fig. 40. Some of the apparent confusion surrounding the values for the various P's can be attributed to simple experimental scatter in the data.

As for density changes produced by temperature variations alone, the half-width of concentration profiles is larger than for velocity, and the profiles themselves are consequently fuller, as shown in Fig. 41. Indeed, nondimensional temperature and concentration profiles have been found to be virtually identical, as shown in Fig. 42, which uses the data of Refs. 42 and 55.

Introducing a turbulent mass diffusion coefficient D_T through

$$\rho D_T \frac{\partial C}{\partial y} = -\rho \overline{vc} \tag{51}$$

we may express this quantity in relation to turbulent momentum or heat transfer through a turbulent Schmidt number

$$Sc_T \equiv \nu_T / D_T \tag{52}$$

or a turbulent Lewis number

$$Le_T \equiv \frac{D_T}{\kappa_T / \rho C_p} = \frac{Pr_T}{Sc_T} \tag{53}$$

Most workers agree that for the axisymmetric case $Pr_T \approx Sc_T \approx 0.7$, which leads to $Le_T \approx 1.0$. A radial variation of Sc_T presented in Ref. 44 is shown as Fig. 43.

Again here, there are less data for planar cases. Reference 46 presents velocity and concentration profiles for $U_\infty = 0$, and these are given in Fig. 44. Using this and data from planar heated wake flows, Reynolds[44] and others suggest that, in this geometry, $Pr_T \approx Sc_T \approx 0.5$. This difference for planar and axisymmetric cases also has not been explained adequately.

Finally, for mean-flow quantities the rate of "entrainment" of external stream fluid into the main mixing zone is of interest for engineering analysis and some calculation schemes. Measurements of that quantity have been made in Refs. 56 and 57 for axisymmetric cases with $U_\infty = 0$. It has been found that the total mass-flow rate in the mixing zone divided by the mass-flow rate of the injectant, $\dot{m}/\dot{m}_j$, is a simple function of the quantity $(\rho_\infty/\rho_j)^{1/2}(x/d)$, as shown in Fig. 45.

The accurate measurement of flow quantities involving concentration fluctuations is generally conceded to be a difficult task, and various methods have been used (e.g., Refs. 58-60). The experiments of Ref. 58 were for the axisymmetric geometry with $U_\infty = 0$, and the variable composition was provided by oil smoke mixed with the air in the injectant. The measurements were made using the light-scatter technique. The data of Ref. 59 were for a round helium jet into air with $U_\infty = 0$, and the measurements were obtained with a combined hot-wire/hot-film sensor.

The technique of Raman scattering of laser light was used in Ref. 60 to obtain detailed mean and turbulence measurements in a round natural gas (95% methane) jet exhausting into stationary air. Radial profiles of the rms of the normalized concentration fluctuations at various x/d are shown in Fig. 46. Also shown for comparison are the results of Becker et al.[58] These profiles are in close agreement with the corresponding measurements for temperature in Ref. 61. The axial variation of the intensity of the concentration fluctuations is shown in Fig. 47, along with corresponding results for the velocity field and the temperature field from Ref. 61 and the concentration field from Ref. 58. The axial variation of the integral scale, $\Lambda_{c/d}$, is shown in Fig. 48, along with the results of Ref. 58. These results can be represented by

$$\Lambda_c/d = 0.0445\,(x/d) \tag{54}$$

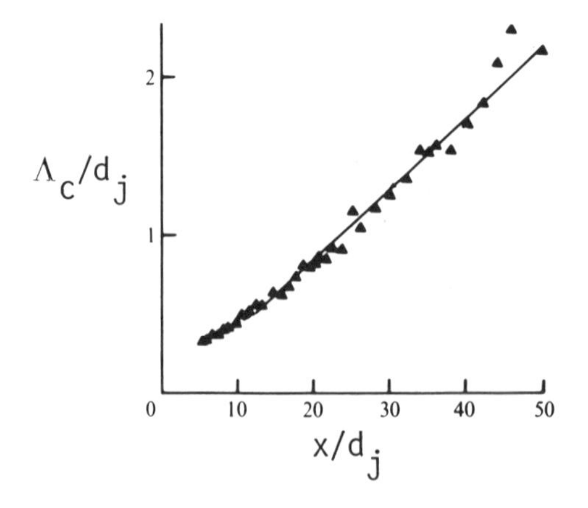

Fig. 48 Streamwise variation of macro length scale for an axisymmetric methane jet into stationary air[60]; ——— Ref. 58 (published with permission).

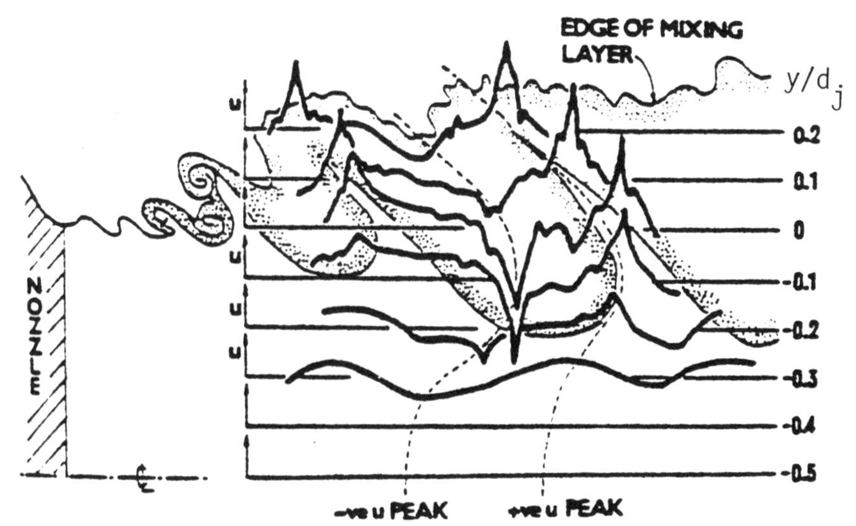

Fig. 49 Velocity (u) time histories at $x/d_j = 2$ in a turbulent water jet measured with an array of hot wires[67] (published with permission).

This linear variation can be reinterpreted in terms of the Eulerian time scale to infer the effects of mean density variations.[60]

6. Large-Scale, Orderly Structure in Jets

Within the past several years, a group of experiments employing various flow visualization techniques[62-64] has shown what appears to be a very orderly, large-scale structure in the turbulence in planar mixing layers. It is, perhaps, noteworthy that all of these experiments were conducted at what amounts to a low Reynolds number for the flow conditions (Mach number, etc.). Chandrsuda and Bradshaw[65] have conducted a thorough study and found that the structures seen were primarily remnants of the transition process, possibly accentuated by the flow visualization techniques themselves. They conclude, in general, "that an asymptotic state with a well-defined large eddy structure exists, but that it is not spectacularly more orderly than that of other shear layers." A closely related flowfield to the planar mixing layer is that in the first few diameters of an axisymmetric jet. A similar large-scale structure has been found in that region in Ref. 66. They found that the flow seems to be built around an array of vortices, possibly toroidal, spaced about 1.25 diam apart and moving with a constant speed. A simple analytical model based on vortices of this type was able to reproduce the major observed features of the flow, including the "spikey" nature of the hot-wire signals.

An extensive study of the transitional and turbulent flow in round water and air jets over a range of Reynolds numbers was reported in Ref. 67. Simultaneous dye pictures and velocity-time measurements such as shown superimposed in Fig. 49 for $x/d = 2$ and $Re_d = 50{,}000$ indicate the relationship between features of the velocity-time signals and the shape of the edges of the mixing layer, especially for the inner boundary. It was concluded on the basis of those studies, as well as those of others, that "coherent three-dimensional eddies are such dominating components of the turbulent mixing layer that their structures and interactions are essential aspects of future experimental and modeling studies of the round jet." Clearly, further work including a wider variety of flow situations should remain a priority for the future.

C. Analysis

1. *Initial Region*

Analyses of the flow in the initial region, or potential core, are generally aimed only at predicting the length of this region, i.e., the axial station where centerline values of the flow variables begin to depart from their initial values at the jet exit. The analysis presented by Abramovich[5] is typical and perhaps the most complete. The flow in the mixing layer around the periphery of the jet in this region is analyzed as though it were a planar mixing layer between two infinite streams. The growth of the width of the mixing zone and any deflection of the zone as a whole due to flow interactions are analyzed. When the inner boundary of the mixing zone is predicted to intersect the jet centerline, the initial region is presumed to be terminated abruptly. Generally, the boundary layers that form on the inside and the outside of the jet nozzle are neglected, although crude attempts at including these have been presented.

The analysis is based on the use of integral forms of the equations of motion and an assumed shape of the velocity profile from Ref. 1:

$$\frac{U_\infty - U(x,y)}{U_\infty - U_j} = f\left(\frac{y-h}{b(x)}\right) \tag{55}$$

for the planar case, where h is the half-width of the jet nozzle, y is measured out from the tip of the nozzle lip, and $b(x)$ is the width of the mixing layer.

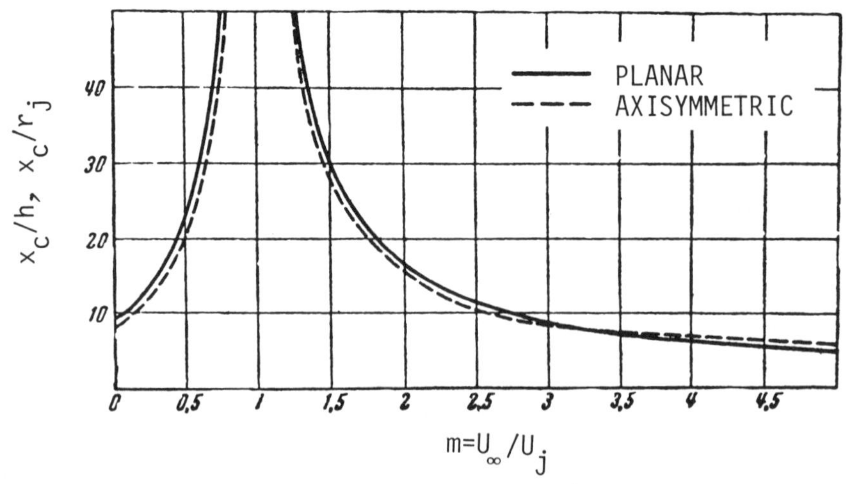

Fig. 50 Analytical prediction of the length of the initial region[5] (published with permission).

The modeling of turbulent transport is accomplished through the experimentally determined width-growth law:

$$b(x) = c \cdot x \tag{56}$$

where $c \approx 0.27$. A similar approach is used for the axisymmetric case. The results are plotted in Fig. 50. Reasonable agreement with experiment is achieved so long as both nozzle boundary layers are small compared to the nozzle half-width. Various extensions for temperature and composition variations are also developed along the same lines in Ref. 5.

2. *Mean-Flow Models*

The oldest and simplest analyses that exist are based on the observed "similarity" of profiles, as was displayed in Figs. 3 and 5. Taking the case of a constant-density round jet with $U_\infty = 0$ as an extremely easy example, one need

Table 5 Eddy viscosity models for main mixing region of jets and wakes

AUTHOR	YEAR	REFERENCE	PLANAR	AXI-SYMMETRIC	VARIABLE DENSITY	EXPRESSION	REMARKS
PRANDTL	1926	1	X	X		$\ell_m^2\left(\frac{\partial u}{\partial y}\right)$	ℓ PROPORTIONAL TO THE WIDTH OF THE MIXING REGION
VON KARMAN	1930	1	X	X		$\kappa^2 \frac{(\partial u/\partial y)^4}{(\partial^2 u/\partial y^2)^2}$	
TAYLOR	1932	1	X	X		$\ell_w^2\left(\frac{\partial u}{\partial y}\right)$	$\ell_w = \sqrt{2}\,\ell_m$
PRANDTL	1942	1	X	X		$\ell_m^2\sqrt{\left(\frac{\partial u}{\partial y}\right)^2 + \ell_1^2\left(\frac{\partial^2 u}{\partial y^2}\right)^2}$	REQUIRES TWO MIXING LENGTHS
PRANDTL	1942	1	X	X		$\kappa_1 b(u_{MAX} - u_{MIN})$	INTRODUCED "VELOCITY DIFFERENCE" CONCEPT; WITH b TAKEN AS $b_{1/2}$, κ_1 = 0.037 IN PLANAR JETS AND κ_1 = 0.25 IN AXI-SYMMETRIC JETS.
SCHLICHTING	1942	1	X			$0.0222\, u_e C_D d$	WAKE OF A CYLINDER OF ARBITRARY CROSS SECTION
CLAUSER	1956	72	X			$\kappa u_e \delta^* = \kappa \int_0^\infty \lvert u_e - u \rvert\, dy$	APPLIED TO "WAKE"-LIKE OUTER REGION OF A BOUNDARY LAYER, $0.016 < \kappa < 0.018$
HINZE	1959	4	X			$0.016\, u_e d$	WAKE OF A CIRCULAR CYLINDER
TING-LIBBY	1960	73		X	X	$\rho^2 \varepsilon = \frac{2\rho_c^2 \varepsilon_o}{y^2}\int_0^y \frac{\rho}{\rho_c}\, y'\, dy'$	ε_o IS THE CONSTANT DENSITY EDDY VISCOSITY AND ρ_c IS THE CENTER-LINE DENSITY
TING-LIBBY	1960	73	X		X	$\rho^2 \varepsilon = \rho_c^2 \varepsilon_o$	ε_o IS THE CONSTANT DENSITY EDDY VISCOSITY AND ρ_c IS THE CENTER-LINE DENSITY
FERRI, ET AL	1962	74		X	X	$\rho\varepsilon = 0.025\left((\rho u)_{MAX} - (\rho u)_{MIN}\right)$	EXTENDED PRANDTL'S THIRD MODEL TO VARIABLE DENSITY, INTRODUCED "MASS FLOW DIFFERENCE" CONCEPT
BLOOM & STEIGER	1963	75		X	X	$\rho\varepsilon = \kappa\delta' \rho_c (u_{MAX} - u_{MIN})$	ATTEMPT TO EXTEND PRANDTL'S THIRD MODEL TO VARIABLE DENSITY, δ' IS TRANSFORMED WAKE RADIUS
SCHETZ	1963	76	X		X	$\rho^2\varepsilon = 0.037\rho_c\left((\rho u)_{MAX} - (\rho u)_{MIN}\right)$	SIMPLE APPLICATION OF "MASS FLOW DIFFERENCE" TO PLANAR FLOWS
ALPINIERI	1964	53		X	X	$\frac{\rho\varepsilon}{\rho_i u_i} = 0.025\, b_{1/2}\left(\frac{\rho_e u_c}{\rho_i u_i} + \frac{\rho_e u_e^2}{\rho_i u_i^2}\right)$	
ZAKKAY, ET AL	1964	48		X	X	$0.011\, b_{1/2} u_c$	PRESUMES THAT CENTERLINE VELOCITY AND CONCENTRATION DECAY AS x^{-2}
SCHETZ	1968	77	X	X	X	"TURBULENT VISCOSITY PROPORTIONAL TO MASS FLOW DEFECT (OR EXCESS) IN THE MIXING REGION"	UNIFIED MODEL

only assert that the integral of the streamwise momentum remains constant:

$$J \equiv \rho \int U^2(x,r)\,dA = \text{const} \tag{57}$$

and that $b(x) \sim x$ from experiment to proceed. Since the profile is taken as similar,

$$\frac{U(x,r)}{U(x,0)} = f\left(\frac{r}{b}\right) \tag{58}$$

Substituting into Eq. (57), one gets

$$U(x,0) = \text{const}\,(1/b)\sqrt{J/\rho} \tag{59a}$$

and finally

$$U_c(x) \equiv U(x,0) = \text{const}\,(1/x)\sqrt{J/\rho} \tag{59b}$$

since b is proportional to x. The corresponding result for a planar jet with $U_\infty = 0$ is $U_c \sim 1/\sqrt{x}$. The analyses can and have been carried further to produce complete solutions for the velocity profiles within an unknown constant by Tollmein[68] and Görtler.[69] The unknown constant must be determined by experiment, and then good agreement with the shape of experimental profiles is obtained. Two comments about this class of solutions are in order in the present context. First, these solutions are not easily extended to cases with $U_\infty \neq 0$. Second, there is some doubt that the basic assumption in Eq. (57) need always be true.[70]

Over the years, various approximate methods for treating jet mixing problems with $U_\infty \neq 0$ have been presented (see Refs. 5 and 71). These methods were generally concerned primarily with overcoming the difficulties of solving the equations of motion. The widespread availability of large digital computers that permit numerically "exact" solutions has rendered these methods largely obsolete, and we shall not cover them here.

Since it has now been assumed that the equations of motion can be solved, within the limits of mean-flow turbulence models, the entire discussion reduces to a choice of an eddy viscosity [or mixing length; see Eq. (14)] model. The major models that have been used are listed here in Table 5. It is interesting to observe that the development of new eddy viscosity models has dropped off sharply in recent years. This is obviously a result of a diversion of interest to the newer, higher-order models. This does not mean, however, that useful predictions cannot be obtained using eddy viscosity (or mixing length) models, as we shall see below.

Looking at Table 5, it is important to note that the three entries following Prandtl's third model (which is the most widely used of the first five in Table 5) can be shown to be equivalent to it and to each other. Consider first the Clauser model

$$\nu_T = K_l U_\infty \delta^* \tag{60}$$

where $K_l \approx 0.018$ and where δ^* must be interpreted to be based on the absolute value of $[1-(U/U_\infty)]$ and the Prandtl model, using $b_{1/2}(x)$ as the width:

$$\nu_T = 0.037 b_{1/2}(x)\, |U_{max} - U_{min}| \tag{61}$$

For simple profile shapes such as a rectangular or triangular velocity defect or excess, one finds that the two expressions agree exactly in form and to the extent of 0.036 vs 0.037 as the proportionality constant. The "wake" models of Schlichting and Hinze can be reduced to the same form as the Clauser model. Noting that

$$C_D d = 2\theta \tag{62}$$

and taking a representative value of $C_D = 1.20$ for a circular cylinder in high Reynolds number flow, these expressions become respectively,

$$\nu_T = 0.044 U_\infty \theta \tag{63a}$$

$$\nu_T = 0.027 U_\infty \theta \tag{63b}$$

In the treatment of wake problems, it is common to neglect the factor (U/U_∞) in the definition of θ, since $(U/U_\infty) \approx 1$. This does, however, render

$$\theta \approx \Delta^* \left(= \int_{-\infty}^{\infty} \left| 1 - \frac{U}{U_e} \right| dy \right) \tag{64}$$

so that these formulas can as well be written as

$$\nu_T = 0.044 U_\infty \Delta^* \tag{65a}$$

$$\nu_T = 0.027 U_\infty \Delta^* \tag{65b}$$

The Clauser model written in these terms is

$$\nu_T = 0.018 U_\infty \Delta^* \tag{66}$$

where we have taken the displacement thickness appropriate to a "two-sided," planar free mixing problem rather than the "one-sided" boundary-layer case considered by Clauser. It will be shown below that Eq. (66) provides predictions in good agreement with jet experiments. The question arises as to why the constant for wakes $(U_c/U_\infty < 1)$ is so much larger than that for jets $(U_c/U_\infty > 1)$. Abramovich[5] notes this effect and attributes it to increased turbulence caused by the separated base flow in the wake case. This matter was also considered in Ref. 31, and we shall return to this point in a later section. For our purposes here, however, the important result is that these free mixing eddy viscosity models are all equivalent in functional form.

The six models listed in Table 5 following the wake models are attempts to extend the basic Prandtl model to problems involving significant density variations. The Ting-Libby model results from an attempt to apply transformation theory to turbulent free mixing; it has been shown to be unreliable.[74] Ferri's suggestion of utilizing a mass-flow difference to replace the velocity difference in the Prandtl model has provided predictions of unreliable accuracy for the axisymmetric case.[48] However, when the mass-flow difference was applied to the planar case, a good prediction[76] was achieved. The Alpinieri model is contrary in form to any other model and must be viewed as essentially empirical, qualitatively as well as quantitatively. The simple Zakkay axisymmetric model given in Table 5 is based on the presumption that the asymptotic decay of the centerline velocity and jet fluid concentration behave as the inverse of the square of the streamwise distance. As discussed in Sec. II. C.5, it is true that these quantities do behave as x to some negative exponent, but the value is a function of the jet to freestream mass flux ratio, not simply always 2.0. Zakkay asserts that an extended form of his model can be used for cases with $P_c \neq 2.0$. Examination of Ref. 48, however, reveals that it is necessary to know the decay exponent for both the centerline concentration and the velocity a priori before the extended model can be specified.

Since it was possible to demonstrate some unity of the models for planar, constant-density cases, it was instructive to examine new means for extending these models to the compressible case.[77] As these models are all equivalent, one could begin with any one. Rather than the usual procedure of starting with the Prandtl model, Ref. 77 chose to begin with the Clauser model. It was simple, at least formally, to extend this to varying density, i.e.,

$$\nu_T = 0.018 U_\infty \Delta^* = 0.018 U_\infty \int_{-\infty}^{\infty} \left| 1 - \frac{\rho U}{\rho_\infty U_\infty} \right| \mathrm{d}y \tag{67a}$$

and to show that for simple profiles, such as a triangular velocity defect, this expression is equivalent to the planar mass-flow difference model given in Table 5. Recalling that that model produced predictions in good agreement with experiment, it may be stated that the generalized Clauser model, Eq. (67a), can be viewed as an adequate representation of planar, free mixing flows with or without strong density variations. To support that assertion, we present in Figs. 51 and 52 some comparisons of experimental data[28] and predictions based on Eq. (67a). The predictions were obtained using a computerized, finite-difference solution of the boundary-layer form of the equations of motion.[32] The experiments contained a small axial pressure gradient, and, when that was included, even better agreement was achieved. It should be emphasized here that it was mentioned earlier that the Prandtl model and the generalized Clauser model are essentially equivalent in the planar case, so that the same level of agreement could be obtained with the Prandtl model.

This still left the axisymmetric case without a satisfactory eddy viscosity model, especially for variable-density cases. In Ref. 77, the extended Clauser model was extended further to the axisymmetric geometry as follows. Rewrite Eq. (67a) as

$$\mu_T(x) = \rho \nu_T = K_1 \rho U_\infty \Delta^* \tag{67b}$$

where $\mu_T(x)$ is the turbulent viscosity. This can be read to say, "The turbulent viscosity is proportional to the mass-flow defect (or excess) per unit width in the mixing region." One can carry this statement over into axisymmetric flow

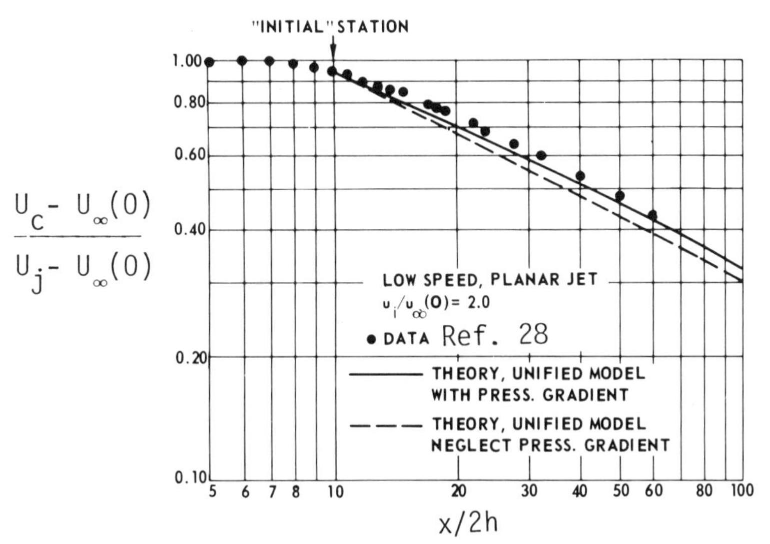

Fig. 51 Comparison of prediction and experiment for centerline velocity variation of a planar jet.[32]

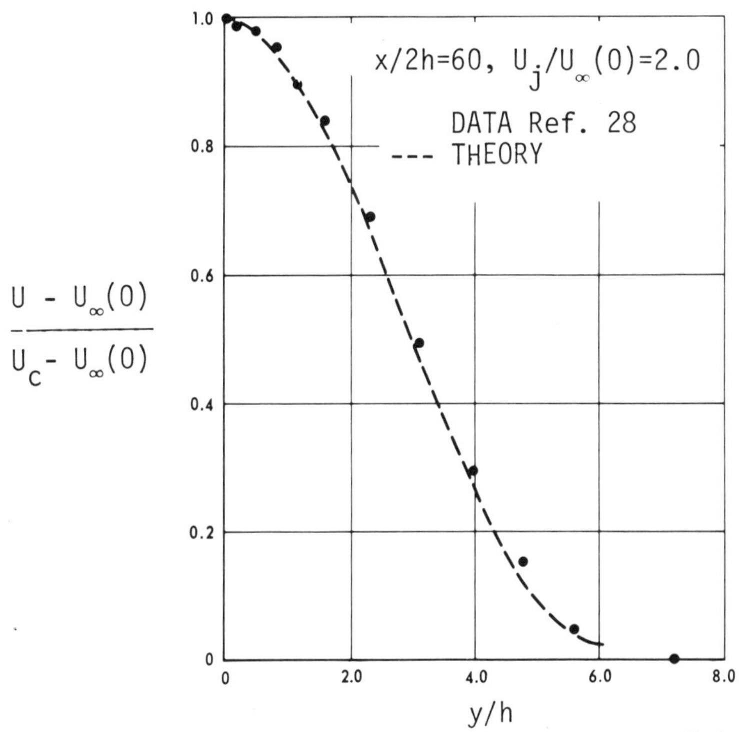

Fig. 52 Comparison of prediction and experiment for transverse velocity profile in a planar jet.[32]

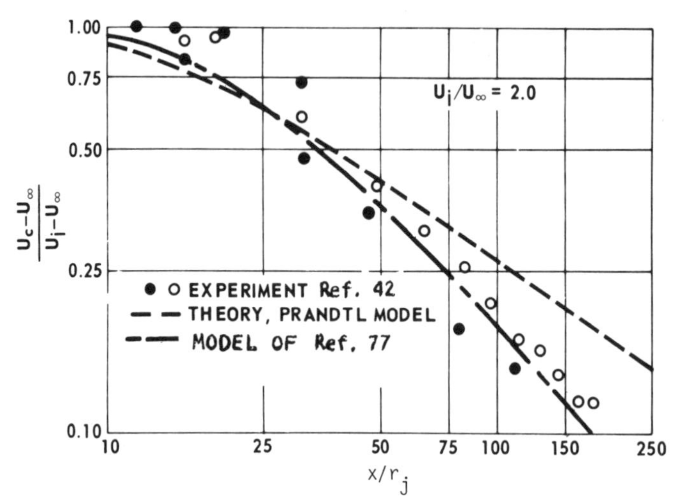

Fig. 53 Comparison of prediction and experiment for centerline velocity variation of an axisymmetric jet.[77]

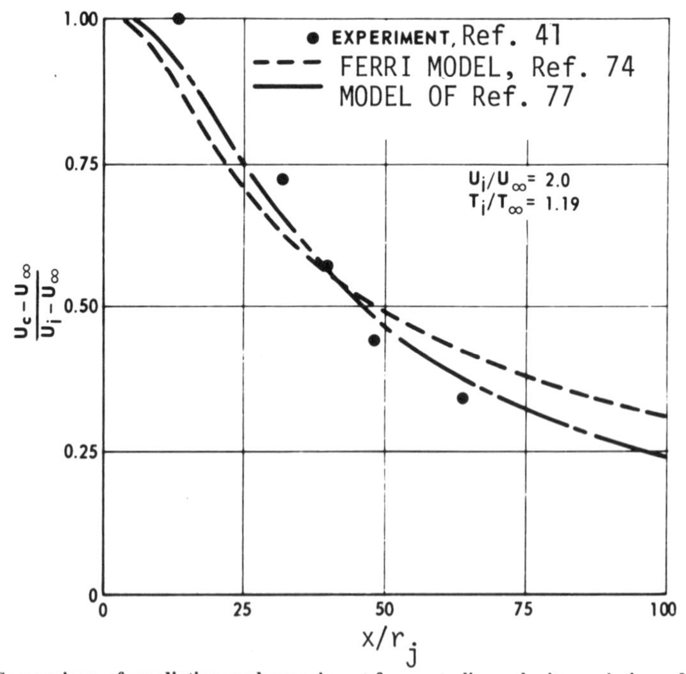

Fig. 54 Comparison of prediction and experiment for centerline velocity variation of a heated, axisymmetric jet.[77]

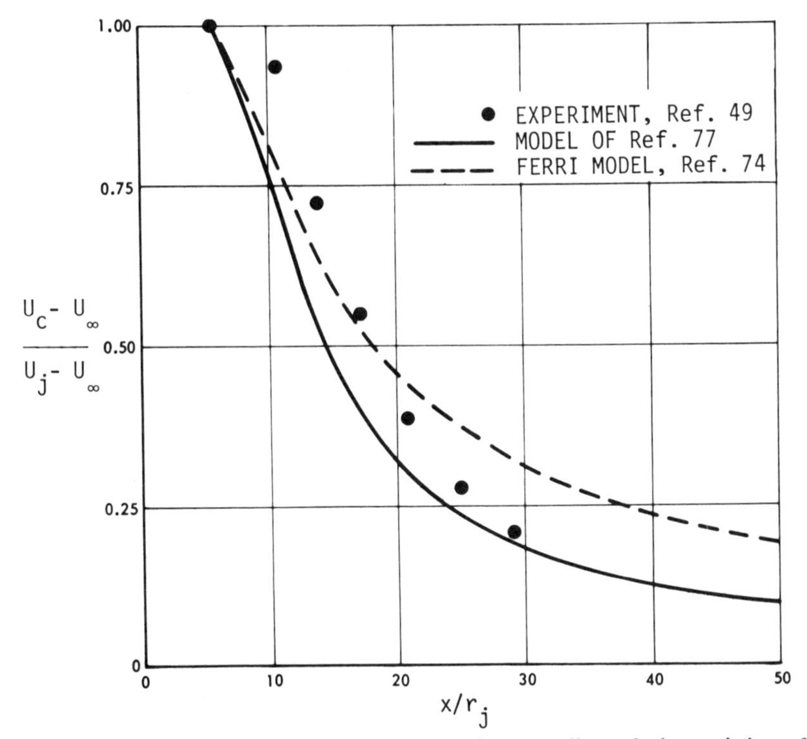

Fig. 55 Comparison of prediction and experiment for centerline velocity variation of an axisymmetric H_2 jet into air.[77]

by defining a new displacement thickness δ_r^* as

$$\pi\rho_\infty U_\infty (\delta_r^*)^2 \equiv \int_0^\infty |\rho_\infty U_\infty - \rho U| 2\pi r \mathrm{d}r \tag{68}$$

This gives

$$\mu_T(x) = K_2 \rho_\infty U_\infty \pi (\delta_r^*)^2 / L \tag{69a}$$

The proportionality constant K_2 had to be determined by a comparison between theory and experiment for one case, as is done with all eddy viscosity models. It is a function of the choice of the characteristic length L; Ref. 77 took $L = r_j$, the initial jet radius. With this, the model becomes

$$\mu_T(x) = (K_2 \pi) \rho_\infty U_\infty (\delta_r^*)^2 / r_j \tag{69b}$$

In order to make the one-time determination of the constant in Eq. (69b) and to begin the comparison of theory and experiment, the essentially constant-density experiments of Ref. 42 were employed. The particular case chosen was for $U_j / U_\infty = 2.0$, and the results are shown in Fig. 53; the value of

$K_2\pi$ determined is 0.018. One can observe that the model of Ref. 77 gives excellent qualitative as well as quantitative agreement with the data; the asymptotic decay predicted by the Prandtl model is at variance with the data.

At this point, comparisons between theory and experiment could be extended to cover a wider range of parameters. The effects of density variations were included, limited first to those due to temperature variations. Obviously, it becomes necessary to employ the thermal energy equation. Consistent with the general boundary-layer approximations employed throughout, this may be written for low-speed cases as

$$C_p\left(\rho U\frac{\partial T}{\partial x}+\rho V\frac{\partial T}{\partial y}\right)=-\frac{1}{y^j}\frac{\partial}{\partial y}\left(y^j C_p \rho \overline{vT'}\right)+U\frac{\mathrm{d}P}{\mathrm{d}x} \tag{70}$$

where $j=0$ for planar and $=1$ for axisymmetric flows. The term on the right-hand side is modeled using Eqs. (48) and (49). Ferri's extension of Prandtl's model was employed along with the model proposed in Ref. 77, and the experimental data of Ref. 41 were used as the standard against which the adequacy of the theories was measured. A constant Prandtl number of 0.75 was used throughout. Results for heated jets ($T_j/T_\infty=1.19$) with $U_j/U_\infty=2.0$ are given in Fig. 54. In these cases also, the solution including the temperature field based on the model of Ref. 77 is generally in better agreement with the data than that based on Prandtl's model or Ferri's extension thereof. This conclusion was reached on the basis of the slope of the centerline velocity and temperature decays.

The last variable that must be considered in the general treatment of this flow problem is jet and freestream composition. Now, we must add a species conservation equation taken as

$$\rho U\frac{\partial C}{\partial x}+\rho V\frac{\partial C}{\partial y}=-\frac{1}{y^j}\frac{\partial}{\partial y}\left(y^j \rho \overline{vc}\right) \tag{71}$$

The term on the right-hand side is modeled using Eqs. (51) and (52). The case of a jet of hydrogen injected into an airstream provides a stringent test of the theory, since there is a very large density gradient across the mixing zone. The tests of Ref. 49 were used to make the comparisons of theory and experiment, and the Ferri model and the model of Ref. 77 were both used with a constant turbulent Schmidt number of 0.75. The particular case from Ref. 49 with $\rho_j U_j/\rho_\infty U_\infty=0.56$, $U_j/U_\infty=6.3$, is show in Figs. 55 and 56. The calculation was started with experimental profiles at $x/r_j=5.9$. Here, as has been generally true, the model of Ref. 77 produced a superior prediction. Other cases with $\rho_j U_j/\rho_\infty U_\infty$ as low as 0.04 have also been considered, with generally good results.

In addition to the prediction of the behavior of centerline values, the adequacy of the prediction of transverse profiles is also of interest. A comparison of prediction and experiment is given as Fig. 57. Again, good agreement is obtained using the eddy viscosity model of Ref. 77.

One further aspect of the specification of an eddy viscosity model is of interest. The Prandtl model and all its descendants, including that introduced

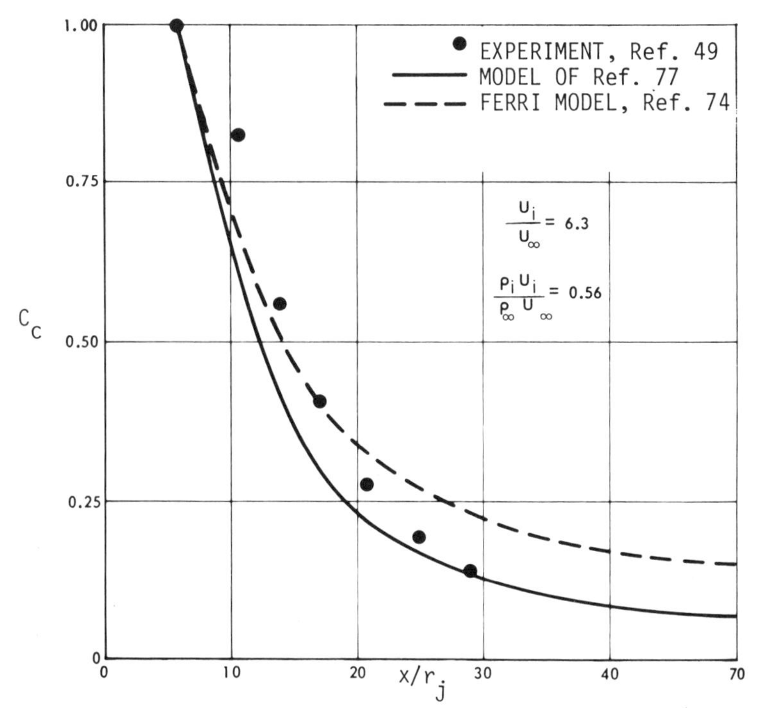

Fig. 56 Comparison of prediction and experiment for centerline concentration decay for an axisymmetric H_2 jet into air.[77]

in Ref. 77, are based upon the difference of some quantity across the mixing region. When the particular difference upon which a given model is based becomes zero, that model will predict a vanishing eddy viscosity. This, of course, is not in agreement with actual experience, and one must simply accept this limitation upon the applicability of the given model. The actual extent of that limitation is, however, important. In the case of the models of Eqs. (67a) and (69b), if the relevant integral of the mass-flow difference becomes zero, the limitation has been reached. The case of uniform initial mass flow (i.e., $\rho_j U_j / \rho_\infty U_\infty = 1.0$) will certainly produce this result, so that the extent of the limitation of these models can be discussed in terms of the initial mass-flow ratio. An axisymmetric case with $\rho_j U_j / \rho_\infty U_\infty$ as high as 0.63 has been considered with a good result, so that the limiting value of $\rho_j U_j / \rho_\infty U_\infty$ must lie closer to unity.

3. *Algebraic Turbulence Function Model*

Aside from the philosophical objection that the mean-flow models (eddy viscosity or mixing length) do not directly reflect the actual turbulent nature of the flow, real cases occur where that deficiency becomes of practical importance. One such instance has been alluded to before. Eddy viscosity models that have been successfully applied to jet or wake cases often do not perform

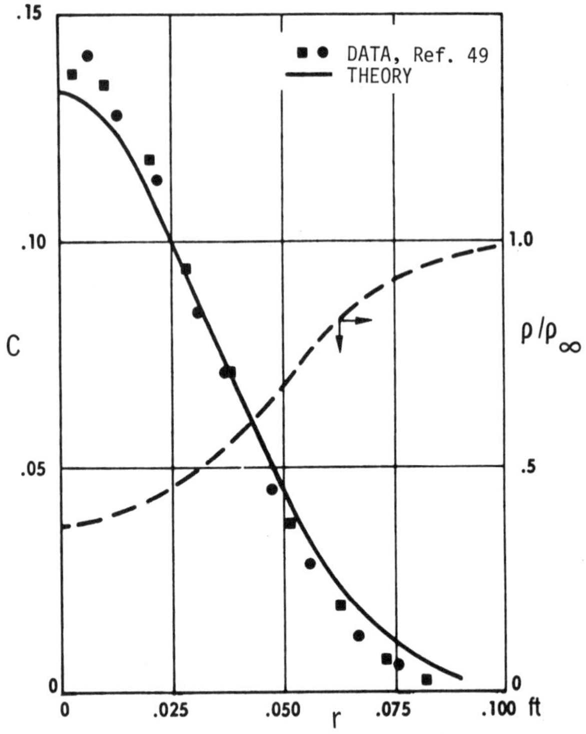

Fig. 57 Comparison of prediction and experiment for a radial concentration profile in an axisymmetric H_2 jet into air.[85]

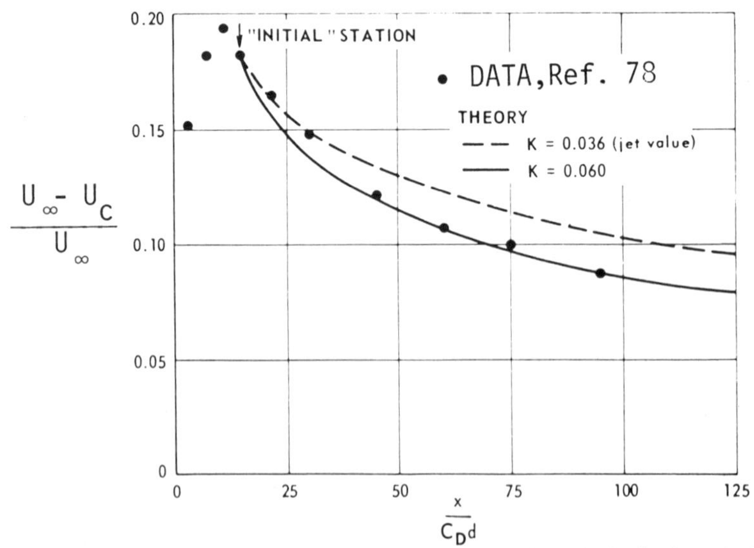

Fig. 58 Comparison of prediction and experiment for centerline velocity in the wake behind a circular cylinder.[32]

adequately when applied to the other case. This is so even when the functional forms of both kinds of models are essentially identical. The discrepancy is primarily centered on the value of the proportionality constant. Compare the "wake" models of Eqs. (65a) and (65b) with the "jet" model Eq. (66). The difference is not simply academic. We have shown that Eq. (66) is capable of good predictions of jet mixing flows (Figs. 51 and 52). What happens if one uses Eq. (66) for a wake case? Figure 58 shows such a comparison for the wake behind a circular cylinder using the data of Ref. 78. Good agreement with experiment can be achieved only with an increase in the value of the proportionality constant significantly above that successfully used for jet problems.

In Ref. 32, the possible simple connection between the observed "turbulence" in various flow problems and the proportionality constant in eddy viscosity models was investigated. Using the planar, constant-density case as an example, one starts with Eq. (12) and approximates the velocity gradient crudely as $(\Delta U)/b$. Replacing the numerical value of the proportionality constant in Eq. (66) by K_3 and using the preceding approximation, we have

$$-\overline{uv}=K_3\left(\frac{\Delta U}{b}\right)\int_{-\infty}^{\infty}|U_\infty-U|\,dy \tag{72}$$

The displacement thickness integral in Eq. (64) can be represented by a constant C_1 times the product of the velocity difference ΔU and the width b. The actual value of C_1 depends upon the specific profile shape, but typical shapes yield a value of roughly one-half. Thus,

$$-\overline{uv}\approx K_3\cdot C_1(\Delta U)^2 \tag{73}$$

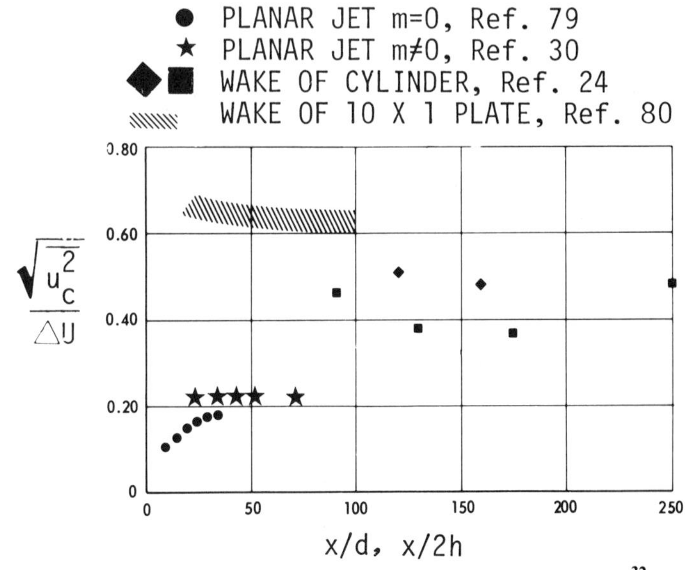

Fig. 59 Axial turbulence intensities for planar mixing flows.[32]

One may write

$$-\overline{uv} = C_2 \sqrt{\overline{u^2}} \sqrt{\overline{v^2}} \tag{74}$$

where C_2 is also about one-half.[1] Also, generally in shear flows

$$u \approx -v \tag{75}$$

so that we finally get

$$\overline{u^2}/(\Delta U)^2 \approx K_3 (C_1/C_2) \tag{76}$$

where (C_1/C_2) is expected to be near unity. Equation (76) achieves the desired result of relating the proportionality constant to a characteristic of the turbulence. Some data for the turbulence/mean-flow information that appears in Eq. (76) have been presented in Fig. 8. More data including planar wakes are shown on Fig. 59. The fact that this quantity is more or less constant for a given flow problem, except in the near-wake or potential core, is comforting in light of Eqs. (66) and (76), since K_3 is presumed to be a constant. Comparing the results for the planar jet with $U_\infty \neq 0$ with those for the wake behind a circular cylinder, we see that the increase of K_3 by a factor of 5/3 required in going from a jet to a wake behind a circular cylinder, as found for the case in Fig. 58, is roughly predicted by Eq. (76).

With all of this, the first-order influence of turbulence structure can be crudely incorporated into an eddy viscosity (or mixing length) model, so long as information of the type in Fig. 59 is available for the general flow problem under consideration.

4. *One-Equation Models*

We come now finally to discuss in some detail the first type of model where the real turbulent character of the fluid enters into the analysis in an explicit way. As stated in Sec. I.C, when a single variable is chosen to represent the turbulence, this is the turbulent kinetic energy (TKE) [see Eq. (16)], as originally suggested by Prandtl and Kolmogorov. There are solid practical, in addition to the obvious philosophical, reasons for increasing the complexity (and thus cost) of mixing analyses to include a treatment of the turbulence. First, we have seen in the previous section that mean-flow information is just not sufficient to describe turbulent shear flows in any general way even for quite simple situations. The addition of such an important new parameter as external stream turbulence level serves to make the case clearer still. Second, the gradient transport model, Eqs. (11) and (12), used as the basis of the eddy viscosity and mixing length models is unrealistic when $\partial U/\partial y \rightarrow 0$. Third, transport processes in shear flows are, in general, influenced by upstream turbulent processes as well as local happenings, and the mean-flow models are all exclusively local in nature.

In order to implement a turbulent kinetic energy approach, each of the terms on the right-hand side of Eq. (17) (or its equivalent in other geometries) must be modeled in the same sense that $\overline{uv}$ had to be modeled in Eq. (8) via, for example, Eq. (11) or (12). That is, these terms must be related to the mean-flow variables (U, V), the turbulent shear $-\rho\overline{uv}$, and/or the turbulence kinetic energy k. This is because we have only three equations, Eqs. (7, 8, and 17), to determine the three unknowns (U, V, k). Indeed, we see that $\overline{uv}$ must still be related to some combination of U, V, and k.

The modeling of the last term on the right-hand side of Eq. (17) is generally viewed as noncontroversial. Since viscous dissipation of turbulent energy takes place predominantly at the smaller eddy sizes and these scales have been found to be nearly locally isotropic, the exact result for dissipation under those conditions is generally simply carried over directly, i.e.,

$$-\mu\overline{\sum\left(\frac{\partial u_i}{\partial x_j}\right)^2} \to C_D\frac{\rho k^{3/2}}{l} \tag{77}$$

where C_D is an empirical constant of order 1/10.

Two distinctly different approaches have been proposed for the modeling of the second term on the right-hand side of Eq. (17) and, simultaneously, the relation of $\overline{uv}$ to k. Prandtl, Kolmogorov, and others reinvoked the eddy viscosity model [Eq. (12)] but introduced k through Eq. (18). This is the approach pursued by the Spalding group[81] and most other workers for free mixing problems. The second general type of model was introduced by Bradshaw[25] as Eq. (19) for boundary-layer (i.e, wall-dominated) flows. Harsha[82] has, however, pursued this approach for free mixing flows. Rotta[83] has considered these two general approaches. Equation (18) is criticized, since it is based upon the gradient transport model with its severe limitations, and Eq. (19) is noted to fail whenever the shear stress changes sign. Rotta also derived the interesting result that both models are contained within the equation for the Reynolds stress, Eq. (21). We shall show comparisons with data of the predictions of both types of models.

The first term on the right-hand side of Eq. (17) is generally looked upon as "diffusion." By crude analogy with laminar diffusion or, for that matter, with the gradient transport model for the diffusion of momentum, Eq. (12), workers who use Eq. (18) employ

$$-(\rho\overline{vk}+\overline{vp}) \to \text{const} \times \rho\sqrt{k}l\frac{\partial k}{\partial y} = \frac{\rho\nu_T}{\sigma_k}\frac{\partial k}{\partial y} \tag{78}$$

where σ_k is a Prandtl number for turbulent kinetic energy, ≈ 1. Bradshaw proposed

$$-\overline{vk} = B \cdot k\sqrt{\tau_{\max}/\rho} \tag{79}$$

to be used along with Eq. (19). Harsha, however, used Eq. (19) but with Eq. (78). This term in Eq. (17) is particularly troublesome to model especially since there are no data for the pressure fluctuations out in the flow upon which to base or test a model.

The very active group at Imperial College contributed a comprehensive survey paper[81] to the NASA Conference that formed the basis for Ref. 10, which compared the adequacy of several turbulence models with jet mixing data, all on the same basis. A number of the specific numerical results that follow in this and the next three sections have been taken from that paper. We shall try to be clear as to the source of the material as we proceed. Also, it might be of use to note that the interested reader will find Ref. 84 very helpful in listing the appropriate form of the equations to be used for various turbulent models in different geometries.

In beginning the comparisons of predictions with experiment, consider first a case with $U_j/U_\infty = 4.0$ from Ref. 42. This essentially constant-density case was one of the test cases used in Ref. 9. In Fig. 60, we have the centerline velocity predictions based on the eddy viscosity model, Eq. (69b) from Ref. 85, the mixing length model and the TKE model based upon Eqs. (18) and (78) from Ref. 81, and the TKE model based upon Eqs. (19) and (78) from Ref. 82, all compared with experiment. All of the predictions are in reasonable agreement with data. It appears that the various models would all give predictions even closer together if they were started at the end of the potential core ($x/d \approx 7$) rather than at the station specified for the cases of Ref. 10, which was still in the potential core.

The H_2-air mixing cases of Ref. 49 provide test cases with very large density differences across the mixing layer. The case with $\rho_j U_j/\rho_\infty U_\infty = 0.56$ has been discussed in this volume before, and it was also used as one of the test cases in Ref. 10. Figures 61 and 62 show the centerline variations of velocity and concentration, comparing the same models as for Fig. 60 with the data. These results must be interpreted in light of the numerical values taken for the various turbulent "Prandtl numbers" involved. The choices that were used are listed in Table 6. Again in this case it is hard to detect any clear advantage of either TKE model over the mean-flow models. This is all the more clear when one notes the confusion introduced by various choices of the values in Table 6. Harsha, for example, appears to use one set of values for some of the calculations in Ref. 10 and another set for other calculations.

An important further point concerning TKE models emerges here. With the additional flow variable k and an accompanying differential equation for it, one needs boundary conditions on that variable as well as the usual requirements for (U, V) [and maybe (T, C)]. This is often not a trivial matter. For example, in Refs. 42 and 49, no experimental data for k, or any other turbulence data, are reported. It therefore becomes necessary to "estimate" the values of the initial and boundary conditions. Obviously, this is a perilous undertaking, and the resulting predictions can be expected to be sensitive to the specific choices made. The situation is most difficult with regard to the initial conditions on k. One must interpret the initial profiles for U (and T and C), i.e, mean-flow variables, to generate an initial profile for k. Generally, the boundary condition on k at the outer edge of the mixing region

Table 6 Numerical values for Pr_T, Sc_T, σ_k used in calculations

Reference	Model type	Pr_T	Sc_T	σ_k
85	ν_T	0.75	0.75	. . .
82	TKE	0.85	0.85	0.7
81	l_m	0.5	0.5	. . .
81	TKE	0.5	0.5	0.7

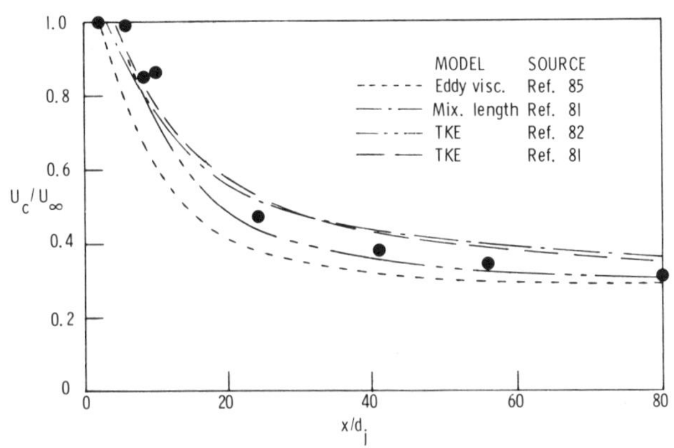

Fig. 60 Comparison of predictions with several models with the axisymmetric jet experiments of Ref. 42[9] (published with permission).

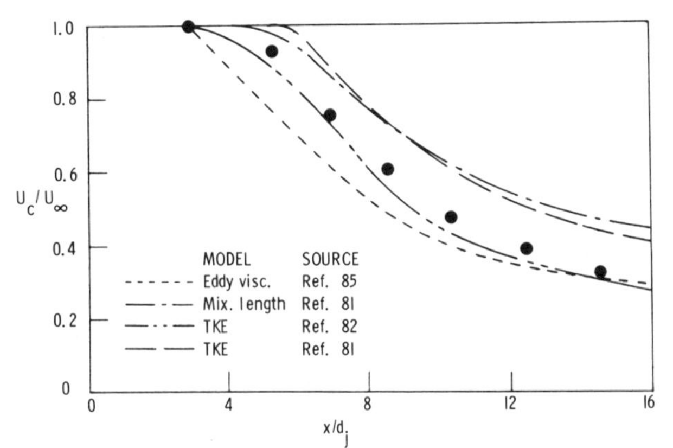

Fig. 61 Comparison of predictions with several models with the axisymmetric H_2 jet into air experiment of Ref. 49: centerline velocity[9] (published with permission).

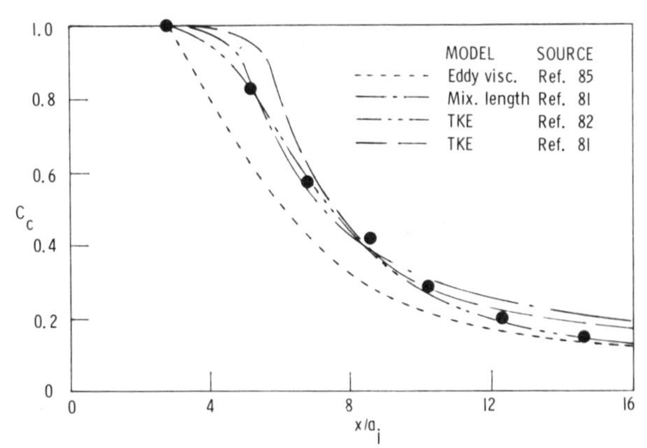

Fig. 62 Comparison of predictions with several models with the axisymmetric H_2 jet into air experiment of Ref. 49: centerline concentration[9] (published with permission).

is taken as zero or the freestream turbulence level, if that is provided. The whole matter is discussed in some detail in Ref. 81. This difficulty is an important limitation on the routine use of TKE models for preliminary design estimates. In that situation, the actual initial and boundary conditions for the mean-flow variables are generally not precisely known and must be estimated themselves. Carrying that process further to produce an initial profile for k is clearly fraught with difficulties.

Lastly, it may be of interest to note here that the TKE models have not been found to provide markedly better predictions than the mean-flow models for wall-dominated flows.[86]

Despite all of this, most workers in the field now view TKE models as superior to the mean-flow models. That conclusion is based upon strong philosophical and somewhat weaker practical reasons. Philosophically, one simply prefers to be treating the turbulent nature of the flow explicitly, albeit perhaps crudely. Practically, the case is not as strong. There are some cases where the turbulence simply must be treated. The high freestream turbulence situation is the most obvious example. Also, looking at a large group of comparisons of predictions and data, an advantage to the TKE is perceived by some observers. This is especially true if wake cases are included in addition to jets.

At this point, we introduce a simplified TKE model that retains all of the philosophical and most of the practical advantages of the full TKE model, but at a great reduction in computational complexity and cost. The approach is an extension of the work of Refs. 87 and 88. The full differential forms of the continuity and momentum (and thermal energy and species conservation, if required) equations are retained. The TKE equation is, however, recast as an ordinary differential equation by integrating across the entire mixing zone:

$$\int_0^b \frac{\partial}{\partial x}(\rho U k) y^j \mathrm{d}y = \int_0^b (-\rho \overline{uv}) \left(\frac{\partial U}{\partial y}\right) y^j \mathrm{d}y - \frac{C_D}{l} \int_0^b \rho k^{3/2} y^j \mathrm{d}y \tag{80a}$$

Equation (77) has been used, but the other terms have not yet been modeled here. It is interesting to note that the troublesome diffusion term [see Eq. (17)] disappears (!) here through the integration operation. Using Eqs. (12) and (18) and taking $l = C_3 b(x)$ as is common, Eq. (80a) can be rearranged to read

$$\frac{\mathrm{d}\nu_T}{\mathrm{d}x} = \frac{C_3^2 U_\infty}{2}\left(\frac{\mathcal{I}_2}{\mathcal{I}_1}\right) - \frac{C_D \nu_T^2}{4U_\infty C_3^2 b^2 \mathcal{I}_1} - \frac{\nu_T}{2\mathcal{I}_1}\frac{\mathrm{d}\mathcal{I}_1}{\mathrm{d}x} + \frac{\nu_T}{2b}\frac{\mathrm{d}b}{\mathrm{d}x} - \frac{\nu_T}{2U_\infty}\frac{\mathrm{d}U_\infty}{\mathrm{d}x} \tag{80b}$$

where

$$\mathcal{I}_1(x) \equiv \int_0^1 \frac{U}{U_\infty}\left(\frac{y}{b}\right)^j \mathrm{d}\left(\frac{y}{b}\right) \tag{81a}$$

$$\mathcal{I}_2(x) \equiv \int_0^1 \left(\frac{\partial(U/U_\infty)}{\partial(y/b)}\right)^2 \left(\frac{y}{b}\right)^j \mathrm{d}\left(\frac{y}{b}\right) \tag{81b}$$

and we have assumed that $\nu_T = \nu_T(x)$ alone and shown only the constant-

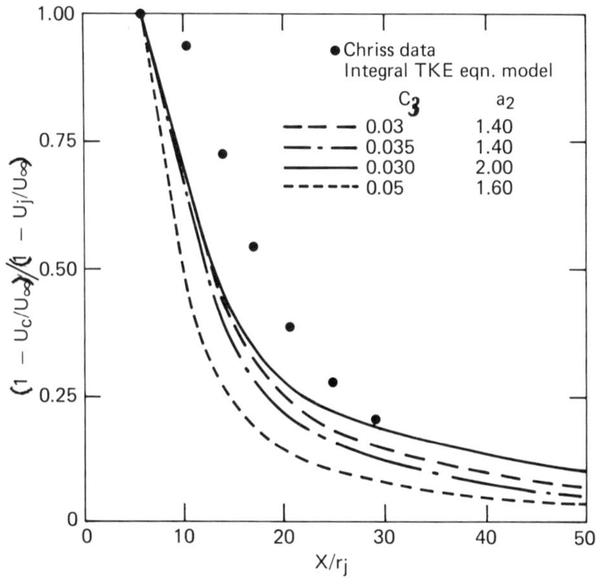

Fig. 63 Comparison of predictions with an integrated TKE model and the experiments of Ref. 49: centerline velocity.

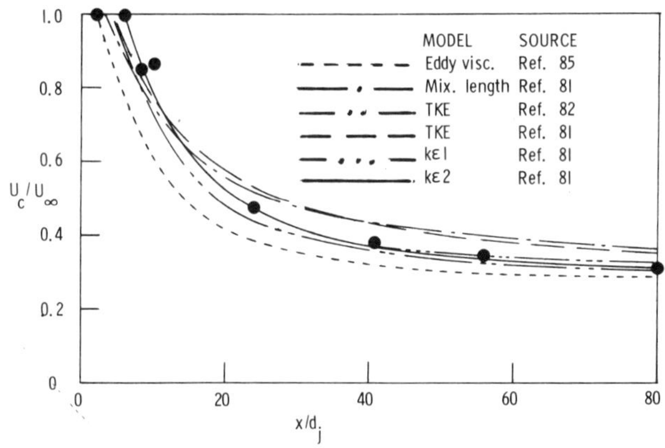

Fig. 64 Comparison of predictions with mean-flow, one-equation, and two-equation models with the axisymmetric jet experiment of Ref. 42[9] (published with permission).

density case as an example. This differential equation can be solved very easily in a streamwise marching fashion as the parabolic partial differential equation system formed by the other equations of motion is solved. All of the terms on the right-hand side can be determined step by step as the whole solution proceeds downstream. The profiles determined at each station are used to evaluate $\mathcal{I}_1$, $\mathcal{I}_2$, and b.

The total mathematical/numerical problem is sharply reduced in scope in this way. This can be emphasized by noting here, as an aside, that Bradshaw's modeling of the TKE equation renders that equation hyperbolic, not parabolic, so that the method of characteristics must be used to solve the partial differential formulation. What, however, has been achieved by this reduction in mathematical complexity? The simplicity of the eddy viscosity formulation is retained, but now we have a differential equation to predict its value, not just an algebraic formula. This means that important upstream history effects or local complex phenomena are now explicitly accounted for in the formulation. The results of a prediction compared with the H_2-air case considered before are shown in Fig. 63 ($a_2/b = C_D/l$). The agreement is about as good as any other model. Since this approach is quite simple and retains most of the advantages of the full TKE model, and since it can easily be incorporated into any mean-flow model computational scheme, it is suggested here as a suitable minimum level in turbulence modeling that should be used in any modern prediction scheme.

5. *Two-Equation Models*

Most of the deficiencies in the performance of the TKE models are attributed to the manner in which the length scale is modeled. That quantity generally appears directly in the modeled form of the various terms, e.g., Eqs. (18, 77, and 78). At the level of the TKE models, the most that can be accommodated is an algebraic formula, usually taken as $l = C_3 b$. It was observed in Ref. 81 that, in the region just beyond the potential core, the mixing rate is predicted as too rapid and, in the far field, the mixing rate is predicted as too slow. This is attributed to the fact that the length scale should apparently be a smaller fraction of the width b in the near field and the opposite for the far field.

In Chapter I, we introduced the Z equation, Eq. (20), which governs the dynamics of the quantity $Z \equiv k^{\alpha} l^{\beta}$. This equation is to be solved in addition to the TKE equation so that, in effect, a differential equation for l has been added. It is, of course, possible to take $\alpha = 0$, $\beta = 1$, so that $Z \equiv l$ directly and to use Eq. (20) in that form to find l. Most workers have not adopted that approach mostly because of problems with modeling the "diffusion" term in Eq. (20). The concept of "diffusion of length scale" has proven hard to accept, when faced directly. Spalding[89] has presented an integrated form of the equation for l_m that was first suggested in Ref. 90. Here again, the integration operation has the pleasant result of eliminating the diffusion term altogether:

$$U_{\mathrm{av}}(x) \frac{\mathrm{d}l_m}{\mathrm{d}x} = a_D k^{1/2} - a_S (\tau_T/\rho)^2 k^{-3/2} \tag{82}$$

where U_{av} is the average velocity across the layer, and $a_D = 0.27$ and $a_S = 0.08$. Of course, τ_T must still be modeled, and this is generally done via Eqs. (12) and (18). This formulation has been used to study the relationship between the spreading rates of various simple mixing flows,[89] but it has not been applied directly to generate complete predictions of the flowfield that can be compared with data for jet mixing problems.

The most active proponents by far of the two-equation models have been the Imperial College group, and they have concentrated on the choice $\alpha = 3/2$,

$\beta = -1$, giving $Z = k^{3/2}/l$, which is proportional to the dissipation rate ϵ [see Eq. (77)]. These models are, therefore, often termed "$k\epsilon$ models." Each term on the right-hand side of the Z (now ϵ) equation still had to be modeled. This was accomplished largely by analogy with the modeling of Reynolds stress and the TKE equation. For example, diffusion of Z is written as

$$-\rho\overline{vz} \sim \rho\sqrt{k}l\frac{\partial Z}{\partial y} \tag{83}$$

There are genuine conceptual problems here. If one looks closely (perhaps too closely) at the various terms of the Z equation, he is really modeling terms that can be described as "dissipation of dissipation" as a particularly undigestible example. In spite of such philosophical problems, these approaches are aimed at solving a real problem, specification of the length scale, and they do offer some practical advantages.

In Ref. 81, the equation suggested for ϵ is

$$\rho U\frac{\partial \epsilon}{\partial x} + \rho V\frac{\partial \epsilon}{\partial y} = \frac{1}{y^j}\frac{\partial}{\partial y}\left(y^j\frac{\rho\nu_T}{\sigma_\epsilon}\frac{\partial \epsilon}{\partial y}\right) + C_{\epsilon 1}\frac{\rho\epsilon}{k}\nu_T\left(\frac{\partial U}{\partial y}\right)^2 - C_{\epsilon 2}\frac{\rho\epsilon^2}{k} \tag{84}$$

with $\sigma_\epsilon = 1.3$, $C_{\epsilon 1} = 1.43$, and $C_{\epsilon 2} = 1.92$ for planar flows. For axisymmetric flows,

$$C_{\epsilon 2} = 1.92 - 0.0667F \tag{85}$$

where

$$F \equiv \left| \frac{b}{2\Delta U}\left(\frac{\mathrm{d}U_c}{\mathrm{d}x} - \left|\frac{\mathrm{d}U_c}{\mathrm{d}x}\right|\right)\right| \tag{86}$$

The TKE equation is retained with $\sigma_k = 1.0$ and

$$\nu_T = C_\mu k^2/\epsilon = (0.09 - 0.04F)k^2/\epsilon \tag{87}$$

This model is called "$k\epsilon 1$." For weak shear flows, i.e., for cases where the velocity defect or excess ΔU is a small fraction of U_∞, an extended version called "$k\epsilon 2$" was presented. This model uses the same equations, but now with $C_{\epsilon 1} = 1.40$, $C_{\epsilon 2} = 1.90$, $\sigma_\epsilon = 1.0$, and

$$C_\mu = 0.09g(\bar{\mathcal{P}}/\epsilon) \tag{88}$$

where

$$\frac{\bar{\mathcal{P}}}{\epsilon} \equiv \int_0^b \tau_T\left(\frac{\mathcal{P}}{\epsilon}\right)y^j\mathrm{d}y \bigg/ \int_0^b \tau_T y^j \mathrm{d}y \tag{89}$$

$\mathcal{P}$ is production of k, and $g(\bar{\mathcal{P}}/\epsilon)$ is given as a graph in Ref. 81. Of course, the types of problems encountered with initial and boundary conditions for the TKE approaches are present, indeed compounded, here.

It might be helpful to the reader to note that there are three types of "ells" commonly used in the literature. The first is the mixing length, which is written as l_m here and defined in Eq. (11). The second is a related quantity at the TKE model level which is defined by Eq. (18), written here as simply l. The third is not explicitly used here, so that we do not really need a special notation for it. This length scale is used, for example, in Ref. 81 (but not Ref. 15) as the length scale in the dissipation term; call it l^* to make its use clear. Equation (77) may now be written either of three ways:

$$\frac{C_D \rho k^{3/2}}{l} = \frac{\rho k^{3/2}}{l^*} = \frac{a_2 \rho k^{3/2}}{b} \tag{90}$$

It then becomes necessary also to modify Eq. (18) to

$$\nu_T = \rho \sqrt{k} l = C_\mu \rho \sqrt{k} l^* \tag{91}$$

Finally, we have $l^* = C \times b$ at the TKE model level. The relationship between these length scales can be seen by looking at a typical example, the planar jet. In that case, $l_m \approx 0.09b$, $l \approx 0.07b$, $l^* \approx 0.875b$, and $C_\mu \approx 0.08$.

Considering the low-speed, axisymmetric jet case of Ref. 42 that was discussed before, comparison of predictions and data are given in Fig. 64. It can be seen that the $k\epsilon 1$ and $k\epsilon 2$ models perform very well.

For the axisymmetric H_2-air cases of Refs. 49 and 91, the situation becomes more confused. In Fig. 65, a comparison of predictions from Ref. 81 with the data of Ref. 91 is shown. Here the TKE and the $k\epsilon 1$ and $k\epsilon 2$ models perform essentially equally. However, for the data of Ref. 49, the relative performance of both $k\epsilon$ models comes out poorer, as shown in Fig. 66.

Looking at the whole of the comparisons of predictions and experiment that are available (Ref. 10 gives an indication of the scope), including again wake cases, many observers see an advantage to the two-equation models. It is known, however, that the use of a set of constants "tuned" to give results for planar jets predict the same spreading rate in axisymmetric jets as for planar jets when experiment indicates that the axisymmetric spreading rate should be significantly less.

6. *Interrelationship Between Some of the Various Models*

In Chapter I, we alluded to the obvious interrelationship between the eddy viscosity and mixing length models. In this section more complex interrelationships between various models will be explored, first for the mean-flow models and then for the one- and two-equation models. The material concerning the mean-flow models is taken from Ref. 92.

The interrelationship between the various types of mean-flow, turbulent transport models can be clearly displayed by considering the integral forms of the boundary-layer equations of mass and momentum. In order to proceed, some information as to typical profile shapes for variations across the layer is also required. Although it is not strictly necessary for that which follows, the assumption of dynamical similarity with axial distance was also made.

The appropriate equations for mass and momentum conservation are

$$\frac{\mathrm{d}}{\mathrm{d}x}\left[\iint_{A_{\text{tot}}} \rho U \mathrm{d}A\right] = e \tag{92}$$

$$\frac{d}{dx}\left[\iint_{A_{tot}} \rho U^2 dA\right] = eU_\infty - \tau_0 N_0 \tag{93}$$

where e is the entrainment and N is the periphery of the control volume. These are the usual forms of these equations where axial pressure gradients have been neglected. Other forms of these integral equations also can be written. In particular, we are interested in the form derived by taking the outer boundary as a stream tube remaining within the viscous mixing region:

$$\frac{d}{dx}\left[\iint_{A_s} \rho U dA\right] = 0 \tag{94}$$

$$\frac{d}{dx}\left[\iint_{A_s} \rho U^2 dA\right] = \tau_s N_s - \tau_0 N_0 \tag{95}$$

With the assumption of dynamical similarity, the velocity profile may be written as

$$(U - U_\infty)/(U_c - U_\infty) = f \tag{96}$$

where f is a function of locally nondimensionalized coordinates, and $f(0) = 1$ and $f(1) = 0$.

Equations (92, 93, and 96) may be solved for the entrainment:

$$\frac{e}{\rho \Delta U A'_T} = \frac{I_1}{2} \frac{(1 + U_e^2/I_2 + 2U_e I_1 - C_{f0} N_0 / 2A'_T I_2)}{(1 + U_e I_1 / 2I_2)} \tag{97}$$

where I_1 through I_4 are integrals of the profile shape (see Ref. 92), and $A'_T = dA_T/dx$ is the axial derivative of the mixing zone area. Also $U_e \equiv U_\infty/(U_c - U_\infty)$. For symmetrical flowfields in the absence of solid boundaries, $C_{f_0} = 0$. Thus, when $U_c \gg U_\infty$,

$$e/\rho \Delta U A'_T = I_1/2 \tag{98}$$

In a similar fashion, Eqs. (92-96) may be solved for the local shear expressed here as a dimensionless coefficient C_f. The result is

$$\frac{C_f N_s}{2A'_T} = \frac{C_{f0} N_0}{2A'_T}(1 - G) + \left(I_2 G - I_4 - \frac{f U_e A_s}{A_T}\right) \tag{99}$$

with

$$G = \frac{I_3(U_e - f) + 2I_4}{I_1 U_e + 2I_2} \tag{100}$$

and $G(0) = 0$, $G(1,x) = 1$. Now μ_T is modeled as

$$\tau_T = \mu_T \frac{\partial U}{\partial n} \tag{101}$$

where n is in the direction perpendicular to the principal direction of motion.

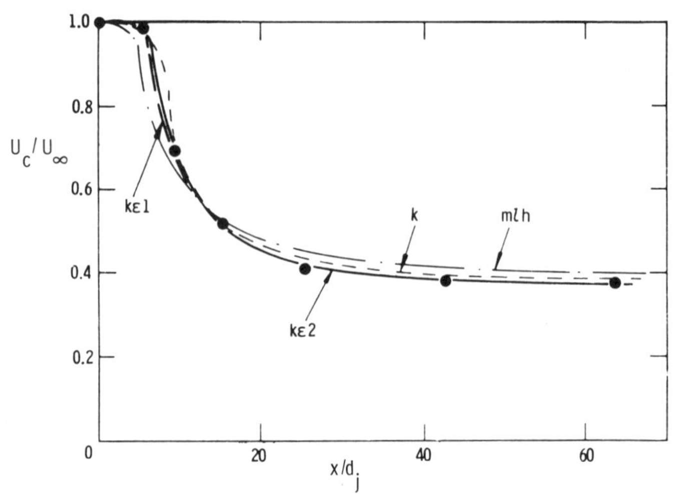

Fig. 65 Comparison of predictions with mean-flow, one-equation, and two-equation models with the axisymmetric H_2 jet into air experiment of Ref. 91: centerline velocity[81] (published with permission).

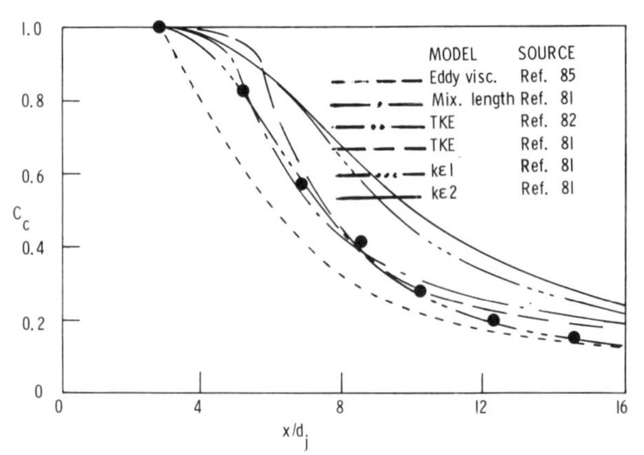

Fig. 66 Comparison of predictions with mean-flow, one-equation, and two-equation models with the axisymmetric H_2 jet into air experiment of Ref. 49[9] (published with permission).

If μ_T is assumed to have a constant value around the periphery of the flow in nonplanar cases, then

$$\frac{\mu_T}{\rho \Delta U A_T'} = \frac{I_2 G - I_4 - f U_e A_s / A_T}{\int_{N_s} \left| \Delta f \right| \mathrm{d}N} \tag{102}$$

where symmetrical flows with no solid boundaries have been presumed.

Equations (97) [or (98)] and (102) can be used to obtain transport models of one type if information of another type is provided. The most direct choice is to take width (or area) growth laws as given and then develop entrainment or eddy viscosity models. This is a somewhat arbitrary choice; however, this selection was made on the basis of the fact that width growth is certainly the easiest quantity to measure.

Let us examine planar and axisymmetric submerged jets ($U_\infty = 0$) and develop entrainment models from width growth information. For the planar case, Eq. (98) gives

$$E \equiv \frac{e}{\rho \Delta U L} = \frac{I_l}{2} \frac{\mathrm{d}b}{\mathrm{d}x} \tag{103}$$

which, with $\mathrm{d}b/\mathrm{d}x = 0.22$ (Ref. 5) and the assumption of either a linear or cosine velocity profile ($I_l = ½$), gives

$$E = 0.055 \tag{104}$$

as entrainment into one side of the jet. This value compares favorably with those obtained directly and independently (0.051 to 0.068) by other workers.[93]

For the axisymmetric case,

$$\frac{e}{\rho (\Delta U)(2\pi b)} = \frac{I_l}{2} \frac{\mathrm{d}b}{\mathrm{d}x} \tag{105}$$

For this case it has been common to introduce an average velocity defined as $U_{av} \equiv U_c I_l$, so that

$$e/\rho U_{av}(2\pi b) = 0.10\text{-}0.12 \tag{106}$$

These values compare well with the values of 0.08 to 0.12 reported elsewhere, e.g., Refs. 56 and 94.

Equation (102) may be used to obtain eddy viscosity models from other types of models for specified flow situations. Examination of Eq. (102) reveals that particularly simple forms of eddy viscosity models will result for flow

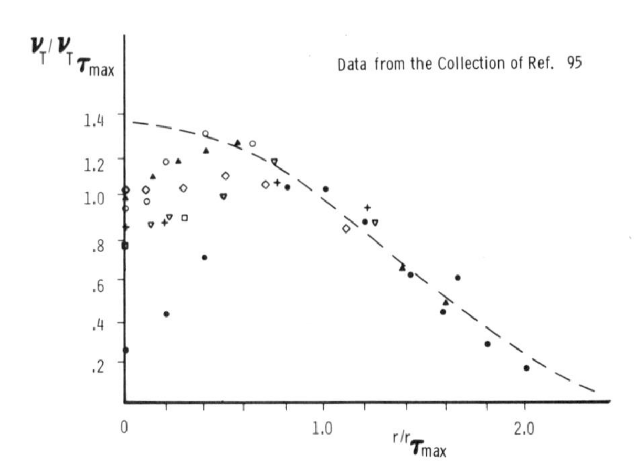

Fig. 67 Comparison of eddy viscosity values derived from measured width growth with data for axisymmetric jets.[92]

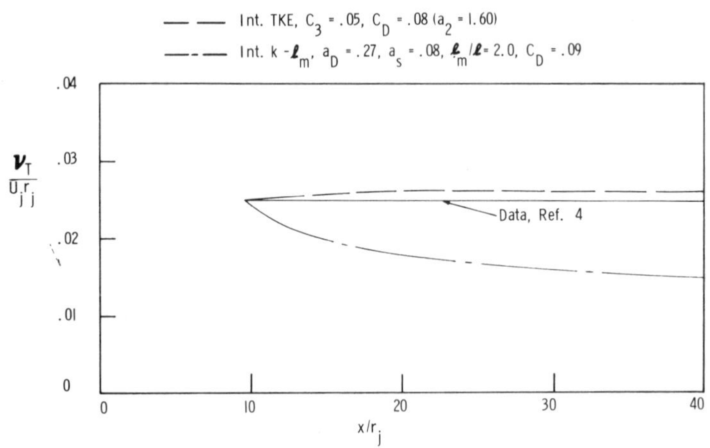

Fig. 68 Eddy viscosity values predicted by integrated one-equation and two-equation models for an axisymmetric jet with $U_\infty = 0$.

situations where the right-hand side is either a constant or a function only of location in the cross plane. In order for the right-hand side of Eq. (102) to be independent of the streamwise coordinate, U_e (and thus G) as well as the integral in the denominator must be x independent. Reference 92 lists several flows, including jets with $U_\infty = 0$, for which these restrictions apply. Consequently, an eddy viscosity model of the form

$$\mu_T/\rho\Delta U A'_T = fn \text{ only of cross-plane location} \tag{107}$$

is adequate for all of these flows, and all x dependence is contained in $\rho(\Delta U)A'_T$.

For the axisymmetric submerged jet, Eq. (102) becomes, with $U_\infty = C_{f_0} = 0$ and $A_T = \pi b^2$,

$$\frac{\mu_T}{\rho\Delta U(\pi b)} = \frac{\mathrm{d}b}{\mathrm{d}x} \frac{I_2 G - I_4}{(\pi/r)\mathrm{d}f/\mathrm{d}r} \tag{108}$$

Results with a simple cosine profile are shown in Fig. 67 compared with data. The radial variation in eddy viscosity is quite good except near the center of the jet where both the numerator and denominator of Eq. (108) approach zero, and the results are very sensitive to profile choice. The important result is that not only is the correct radial variation obtained, but also the complete, correct streamwise dependence is contained in the denominator on the left-hand side of Eq. (108).

The results presented here and others in Ref. 92 can be taken as proof that the common types of mean-flow models, i.e., width growth, entrainment, or eddy viscosity [or mixing length, Eq. (14)], are all directly interrelated. It is interesting to ask now how the one-equation and two-equation models fit into this picture. In order to investigate that question, the axisymmetric jet with $U_\infty = 0$ was selected here as a simple example for study.

For the one-equation (TKE) model, we begin with Eq. (80a). For a case with $U_\infty = 0$, an equation similar to Eq. (80b) results, but with $U_c(x)$ playing the role that U_∞ takes in Eq. (80b). Using the well-accepted width growth law, $b = C_j x = 0.22x$, and a cosine profile shape,[5] the results shown in Fig. 68 are obtained. The calculations were begun at a station, $x/r_j = 9.6$, just downstream of the end of the potential core using an initial value for $\nu_T = 0.025 U_j r_j$ obtained from the material in Ref. 4, p. 425. The values of the required constants C_3 and C_D (or a_2) were taken as 0.05 and 0.08 ($a_2 = 1.60$), as suggested by the information in Ref. 81. The numerical results are sensitive to these choices, and it was seen in Fig. 63 that those choices did not provide the best predictions for that case. However, for the $U_\infty = 0$ case here, this model with those choices of C_3 and C_D does provide a good prediction in comparison with experiment.

For the two-equation models, one has the choice of selecting a definition of Z. The choice $Z \equiv \epsilon$ was made here first. The integrated form of the TKE equation is retained, and an integrated form of Eq. (84) then has to be derived. This is easily done, and Eqs. (12) and (18) and $b = 0.22x$ are used again. The results can be manipulated to give two coupled, ordinary differential equations for $\nu_T(x)$ and $l^*(x)$. Calculations were run with the initial conditions selected in a variety of plausible ways using free jet data. Variations in the values of the required constants also were considered. The results showed a surprising, and disturbing, sensitivity to both types of input information. The predictions generally indicated a growth in $\nu_T(x)$ with streamwise distance which is not shown in the data.

Finally, the use of Eq. (82) with the integrated TKE equation was studied as another example of a two-equation model. The term $U_{av}(x)$ in Eq. (82) was taken simply as $U_c(x)/2$. There is some conceptual difficulty, since Eq. (82) describes $l_m(x)$, and the TKE equation contains $l(x)$. It was necessary to assume a value for the ratio $l_m(x)/l(x)$. Trial values for that ratio between 1.5 and 2.0 were tested. Again, two equations for the two unknowns $\nu_T(x)$ and $l(x)$ [or $l_m(x)$] are obtained. The remainder of the solution, including the setting of initial conditions, was essentially as in the foregoing, but here l_m (initial) was obtained from $l_m = 0.075b$. The predicted results for $\nu_T(x)$ with $l_m(x)/l(x) = 2.0$ are plotted in Fig. 68. The values for ν_T are in reasonable

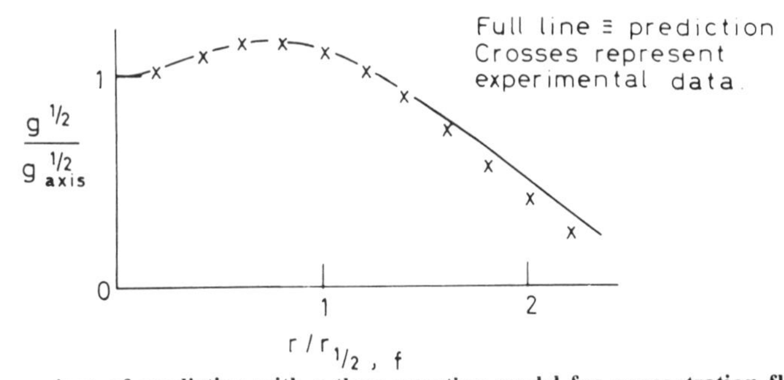

Fig. 69 Comparison of prediction with a three-equation model for concentration fluctuations with the seeded axisymmetric jet experiment of Ref. 58[97] (published with permission).

agreement with experiment, whereas the mixing length was predicted to grow much slower than if l_m/b were constant.

The analysis and results under the last three topics require the use only of overall momentum conservation, a prescribed width growth law ($b \sim x$, for this problem), and the appropriate one or two "turbulent" equations. We saw before that the setting of a width growth law determines the value of the eddy viscosity directly based upon the mean-flow equations alone, Thus, the exercises performed here serve to display, and perhaps amplify, the adequacy and internal consistency of various higher-order models for a simple flow problem where the mean-flow distributions are known ahead of time. Certainly, the inherent problems with initial conditions and the values of the constants involved emerge clearly. This is an application of the "inductive treatment" of Rotta developed earlier and independently in Ref. 96, although the models examined and some of the details of the analysis vary.

7. *Three-Equation Model*

At this point, we return to the presentation of the various differential models in order of increasing complexity. It is of great practical interest to have an analytical description of the variation of the fluctuating character of a scalar (temperature or concentration) in addition to that for the mean values. This is particularly true in problems involving combustion as an example. The full differential two-equation type of model was extended in Ref. 97 to handle this kind of problem. First, one needs an additional "turbulent" equation, this time for the mean-square concentration fluctuation $g \equiv \overline{c^2}$:

$$U\frac{\partial g}{\partial x} + V\frac{\partial g}{\partial y} = -\frac{1}{y^j}\frac{\partial}{\partial y}(y^j\overline{vg}) - 2\overline{vc}\frac{\partial C}{\partial y} - 2\Gamma_c\sum_i\overline{\left(\frac{\partial c}{\partial x_i}\right)^2} \tag{109}$$

The terms on the right-hand side were modeled in what should now be a familiar manner to produce

$$\rho U\frac{\partial g}{\partial x} + \rho V\frac{\partial g}{\partial y} = \frac{1}{y^j}\frac{\partial}{\partial y}\left(y^j\frac{\rho\nu_T}{\sigma_g}\frac{\partial g}{\partial y}\right) + C_{g1}\rho k^{1/2}l\left(\frac{\partial C}{\partial y}\right)^2 - C_{g2}\frac{\rho k^{1/2}g}{l} \tag{110}$$

with $\sigma_g = 0.7$, $C_{gl} = 2.7$, and $C_{g2} = 0.134$.

The TKE equation was retained, but instead of the ϵ equation, an alternate choice for Z, $m=1$, $n=-2$, giving $Z = k/l^2 \equiv W$ was adopted. The coresonding form of the Z equation that was used is

$$\rho U\frac{\partial W}{\partial x} + \rho V\frac{\partial W}{\partial y} = \frac{1}{y^j}\frac{\partial}{\partial y}\left(y^j\frac{\rho\nu_T}{\sigma_W}\frac{\partial W}{\partial y}\right) + C_4\rho k^{1/2}l\left(\frac{\partial^2 U}{\partial y^2}\right)^2$$

$$+ C_6\frac{\rho k^{1/2}}{l}\left(\frac{\partial U}{\partial y}\right)^2 - C_5\frac{\rho k^{1/2}W}{l} \tag{111}$$

with $C_4 = 3.81$, $C_5 = 0.134$, $C_6 = 1.23$, and $\sigma_W = 1.0$. Good agreement with the fluctuating concentration measurements in a round jet with $U_\infty = 0$ reported in Ref. 58 was achieved, as shown in Fig. 69.

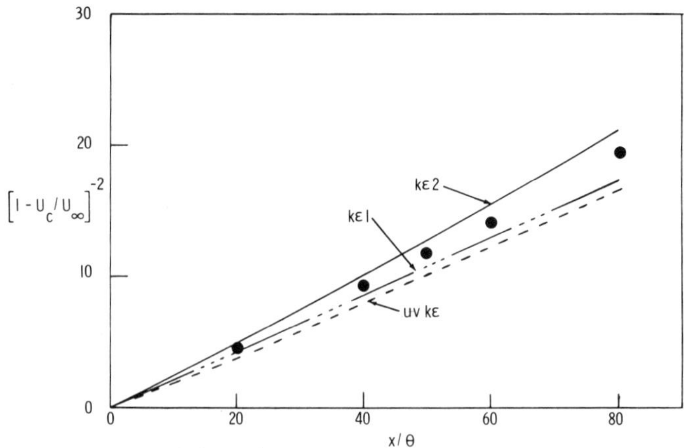

Fig. 70 Comparison of predictions with two-equation and Reynolds stress models for the planar jet experiment of Ref. 30[81] (published with permission).

8. *Reynolds Stress Models*

In order to break away altogether from the concepts of a turbulent viscosity and/or a mixing length, the logical next step in a hierarchy of models is to attempt the treatment of an equation for the Reynolds (turbulent) stress itself, $-\rho\overline{uv}$, Eq. (21). One is then attacking directly the term that caused the central problem at the lowest level of analysis in Eq. (8). Of course, Eq. (21) contains higher-order turbulence terms, e.g., $\overline{uv^2}$, which must now be treated. It is possible to envision writing a new, separate equation for each term of that type. Such equations will, however, necessarily involve terms of the next higher order. The process could be continued, but the whole scheme is unbounded, since the last equation will always contain terms of a higher order. It is necessary, therefore, to "close" the problem by terminating the sequence at some level. At the mean-flow model level, the problem is closed by an eddy viscosity or mixing length model. At the one-equation and two-equation model level, closure is achieved by modeling the higher-order terms in the TKE and Z equations. In like manner, it will now be necessary to model the unknown terms in Eq. (21) for the Reynolds stress. Most workers have also found it desirable to add other equations to the system at this level. Equations for l and/or the normal stresses $\overline{u^2}$, etc., are common choices.

Although it is not the first practical model from a chronological point of view, the model of Hanjalic and Launder[98] follows most simply in logical order from the two-equation models. It is, therefore, presented here first. The "production" term in Eq. (21) is modeled by asserting that $\overline{v^2} \propto k$. The "diffusion" term is modeled through the generalized gradient transport framework, i.e.,

$$-\frac{\partial}{\partial y}\left(\overline{uv^2} + \frac{\overline{pu}}{\rho}\right) \propto -\frac{\partial}{\partial y}\left(k^{1/2} l \frac{\partial \overline{uv}}{\partial y}\right) \tag{112}$$

Note that here a turbulent viscosity ($\mu_T \sim k^{1/2} l$) has crept back into the formulation. The "dissipation" term was neglected with respect to the other terms. Lastly, the "redistribution" term is modeled via two processes: one with turbulent interactions alone, $\sim (k^{1/2}/l)\overline{uv}$, and one which arises via the mean flow gradient, $\sim k\partial U/\partial y$. The final form of the equation for the Reynolds stress becomes

$$U\frac{\partial(\overline{uv})}{\partial x}+V\frac{\partial(\overline{uv})}{\partial y}=\frac{1}{\rho}\frac{\partial}{\partial y}\left(\frac{\mu_T}{\sigma_\tau}\frac{\partial(\overline{uv})}{\partial y}\right)-C_\tau\left(k\frac{\partial U}{\partial y}+\frac{k^{1/2}}{l}(\overline{uv})\right) \tag{113}$$

Since this equation contains k and l, the system is completed with equations for k and l as modeled by the Imperial College group except that τ_T is left as $-\rho\overline{uv}$ as the various terms in Eqs. (17) and (20) are approximated. The details are given in Ref. 81, along with the prediction for the planar jet with $U_\infty = 0$, shown here in Fig. 70. From that comparison and only one other from Ref. 81, no clear advantage over the two-equation models is apparent.

In Ref. 99, a similar level of analysis is presented using equations for $\overline{uv}$, k, and (kl). The modeling of the various terms is, however, quite different in that variations and extensions of Bradshaw's modeling of the diffusion term, Eq. (79) are employed. Good agreement with the author's own experiments is reported.

Rotta[83] concluded, after a careful assessment of the use of a stress equation model for the planar wake problem that "no perceptibly better results could be achieved with the shear stress transport equation than with the Prandtl eddy viscosity relation." Here, he refers to the TKE approach of Prandtl.

At a slightly higher level of mathematical complexity, Donaldson[100] chose to employ an equation for the Reynolds stress, algebraic relations for the length scale l, and separate equations for the normal stresses $\overline{u^2}$, $\overline{v^2}$, $\overline{w^2}$. The details of

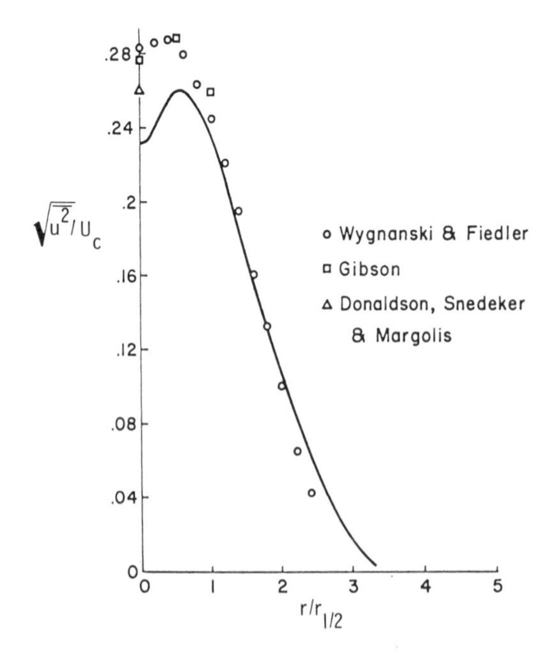

Fig. 71 Comparison of a prediction with a Reynolds stress model and experiment for an axisymmetric jet with $U_\infty = 0$ for axial turbulence intensity[100] (published with permission).

the modeling of all of the terms and some results for an axisymmetric jet with $U_\infty = 0$ are given in Ref. 100. It is worth noting that the turbulent viscosity concept is not invoked in any form. The length scale is, however, not determined from a differential equation. The results for Reynolds and normal stresses are compared with some experimental data in Figs. 71 and 72. Some adjustments to the many constants involved where made to obtain the results shown in these figures, and the reader is referred to Ref. 100 for all of the details. Nonetheless, good results are obviously achievable with this model using tailored constants.

The Harlow-Daly model presented in Ref. 101 employs separate equations for the Reynolds stress, each of the normal stresses, and l (via ϵ). Since each normal stress value is found separately, one does not need an equation for one-half of their sum, i.e., k. Still, this model entails the solution of five equations for turbulent quantities in addition to two (or more) mean-flow equations. The modeling of the terms in the equations for the Reynolds and normal stresses is similar to that used by Hanjalic and Launder. The Harlow-Daly model has been applied by Rodi to the planar jet with $U_\infty = 0$, and the results are compared to experiment in Fig. 73. The mean flow is predicted well, but the individual normal stresses are seen to be predicted poorly. The shear stress and turbulence energy are reported to have been predicted accurately.[15]

A limited comparison of the Hanjalic-Launder and Harlow-Daly models as they treat the triple velocity correlation terms, e.g., $\overline{uv^2}$, is presented in Ref. 102, where it is concluded that the Hanjalic-Launder approach is to be preferred. In Ref. 103, the adequacy of the usual modeling of the "pressure-strain" term [called "redistribution" in Eq. (21)] following Rotta[90] was considered in detail. Many workers have considered the modeling of that term as one weak link in the use of the Reynolds stress equation. It was concluded,

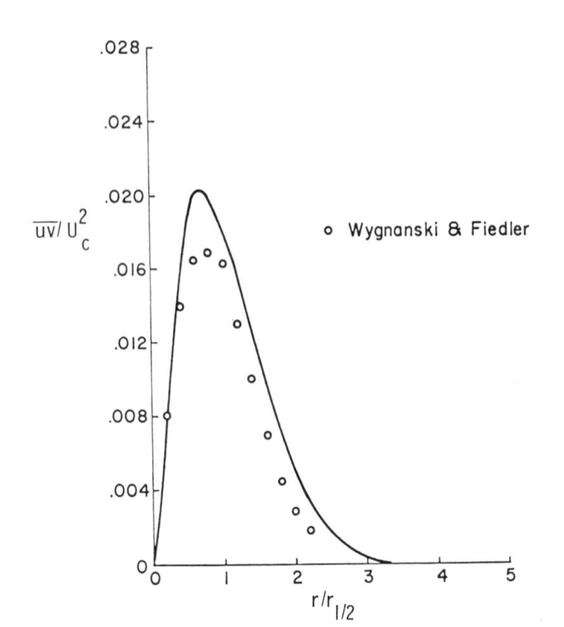

Fig. 72 Comparison of a prediction with a Reynolds stress model and experiment for an axisymmetric jet with $U_\infty = 0$ for Reynolds stress[100] (published with permission).

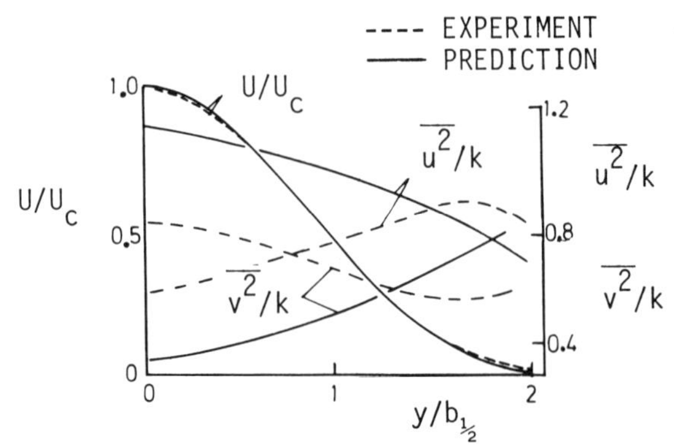

Fig. 73 Comparison of prediction with the Harlow-Daly Reynolds stress model and experiment for a planar jet after Rodi.

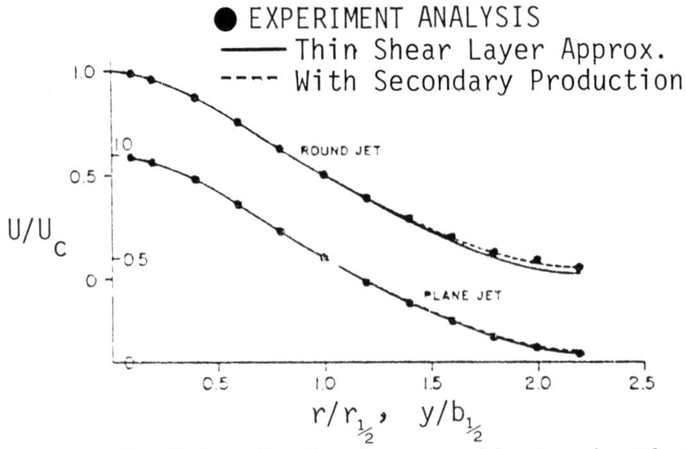

Fig. 74 Comparison of predictions with a Reynolds stress model and experiment for the velocity profiles in planar and axisymmetric jets[104] (published with permission).

however, that no clear case for a basic weakness of the Rotta model could be shown.

For any who came to this point with an overly optimistic outlook, Ref. 104 will make sobering reading. The paper presents the results of extensive trial calculations and comparisons with experiment using the model of Ref. 103 as a baseline model for both planar and axisymmetric geometries. The hope was to develop a more general model that could treat the planar and axisymmetric cases equally well. When numerically accurate solutions based upon the baseline model from Ref. 103 were obtained for axisymmetric cases, it was found that the spreading rate predictions were worse than from the two-equation model, which were in poor agreement with experiment themselves. Next, the inclusion of all of the small extra strain rates generally neglected in thin shear layer calculations was studied at the price of a considerable increase

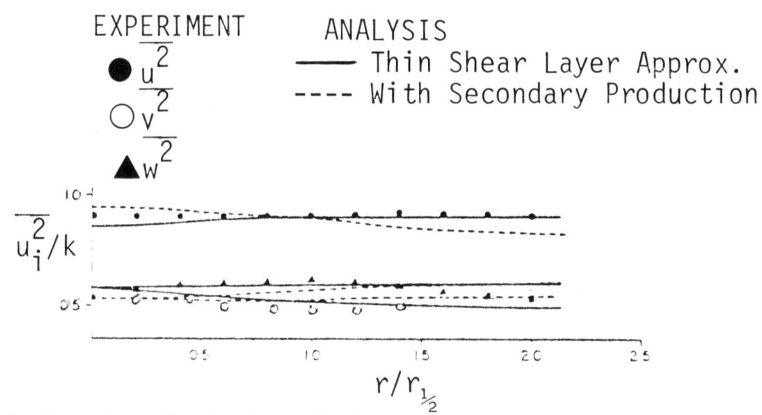

Fig. 75 Comparison of predictions with a Reynolds stress model and experiment for normal stress profiles in an axisymmetric jet[104] (published with permission).

Table 7 Measured and predicted spreading rates

Flow	$db_{1/2}/dx$		Predicted including small terms
Planar jet ($U_\infty = 0$)	0.110	0.123	0.123
Round jet ($U_\infty = 0$)	0.087	0.135	0.136

in computational complexity. It was found, however, that the influence of these terms on spreading rate is, indeed, negligible. These conclusions may be confirmed by reference to Figs. 74 and 75 and Table 7. Even though the normalized profile shape is predicted well, the spreading rate results are not acceptable.

An attempt then was made to find a set of the necessary constants that would perform better for both cases. This was unsuccessful, since the effects of a change in each coefficient were found to be of a similar magnitude in the different cases. Thus, no selective changes that might improve agreement in the axisymetric case while leaving the planar case essentially unaffected seem possible.

The major difficulty at this point in the study seemed to be associated with the modeling of the source terms in the dissipation equation. Further work on swirling flows confirmed that but also indicated problems with the modeling of the pressure strain processes. We shall return to this discussion in Chapter V, which deals with swirling flows in detail. However, it should be stated here that no final solution to these problems is presented in Ref. 104, but some avenues for future study were suggested.

9. Direct Turbulence Models

As stated toward the end of Sec. I.C, the application of models that attempt to simulate the turbulence directly to real flow problems is in its infancy.

Indeed, there are no complete predictions for jet mixing flows available that can be compared to experiment. It is likely that it will be some years before such predictions are available for a wide range of problems.

The "chemical engineering" approach, also briefly described in Sec. I.C, has been applied very approximately to some practical problems, although these do not include the case of interest in this chapter. The reader is referred to Refs. 7 and 105 for a description of some of the available results for cases such as mixing in a closed vessel and mechanically aided mixing.

III. AXIAL PRESSURE GRADIENTS

A. Experimental Studies

1. *Mean-Flow Data*

Many practical jet mixing problems involve mixing in a confined duct, and this generally produces an axial pressure gradient which is intimately coupled to the mixing process, and vice versa. It is, of course, possible to tailor the area distribution of the duct so as to maintain a constant pressure. Two experiments in Ref. 106 were conducted at similar nominal flow conditions except that one had a constant-area and thus variable static pressure duct, and the other had a variable-area duct selected to keep the static pressure constant. The development of the mean flow is shown here in Fig. 76 from that reference.

The general result of high-velocity jet mixing in a duct is increasing static pressure along the flow direction. Indeed, this is the basis for the common "jet ejector" type of fluid pump. However, the "adverse" pressure gradient thus produced, $dP/dx > 0$, can effect a massive rearrangement of the flow, if strong enough. We limit the discussion here to planar or axisymmetric ducts with a jet nozzle of the corresponding geometry on the centerline. Fluid is injected from the nozzle at a velocity U_j, and there is a flow in the surrounding duct at a velocity U_∞ at the injection station. If there is an axial static pressure gradient, the velocity of the surrounding, essentially inviscid flow will be altered through Bernoulli's equation as it proceeds downstream. If we write the surrounding velocity as $U_\delta(x)$, the relationship is

$$U_\delta(x)\frac{dU_\delta(x)}{dx} = -\frac{1}{\rho}\frac{dP}{dx} \tag{114}$$

If $dP/dx > 0$, then $dU_\delta/dx < 0$, and it is possible to reduce $U_\delta(x) \to 0$ for strong enough dP/dx producing a "stalled" flow and substantial recirculation zones in the duct. This is precisely what is observed in experiment. Reference 107 presents the results of a detailed study of such cases, and some of those data are shown here in Figs. 77a-77c for various flow conditions. Figure 77a shows a flow with no recirculation zones, but as the conditions change a small recirculation zone may be seen to be forming in Fig. 77b. Finally, for the conditions in Fig. 77c most of the flowfield is dominated by the recirculation. The tendency toward recirculation can be described by the value of simple parameters involving the total mass flow in the duct, Q, and the pressure-momentum integral $\mathfrak{M}$, which are defined as

$$Q \equiv \int_A \rho U dA \tag{115}$$

$$\mathfrak{M} \equiv \int_A (P+\rho U^2)\,\mathrm{d}A \tag{116}$$

Various workers have combined these in different parameters, H, M, and C_t, which are all simply related algebraically, e.g.,

$$\frac{Q}{\sqrt{2\pi(\mathfrak{M}/\rho)}\;R_0} \equiv H \equiv \frac{1}{\sqrt{2(M+1/2)}} \equiv \frac{1}{\sqrt{1+2/C_t^2}} \tag{117}$$

for the axisymmetric case, where R_0 is the duct radius. C_t was used in Ref. 107 and on Figs. 77a-77c. The static pressure distributions obtained for several values of C_t, including those for Fig. 77, are shown in Fig. 78.

In practice, the recirculation zones just discussed are generally to be avoided, and data for confined jet mixing without recirculation are of most interest. One can determine special conditions where the flow remains "similar" even in the presence of an axial pressure gradient. The condition is that $U_\delta(x)/[U_c(x)-U_\delta(x)]$ remain constant along the duct. Such a case was found and studied in Ref. 108, and it is of some interest especially with regard to the older analyses based on similarity arguments. The recent experiments reported in Ref. 109 for flow in a constant-area duct with the pressure gradient developing freely, but at a level such that recirculation did not develop, produced much useful information. Tests were conducted with the jet velocity held nominally fixed at about 60 m/s and the surrounding flow at the injection station varying from about 15 to 30 m/s. Transverse velocity profiles for the 4:1 velocity ratio test are shown in Fig. 79, where U_m is the average velocity across the duct.

There have been fewer studies of the confined jet in the planar geometry. The reader is referred to Ref. 110 for one such work. A particularly interesting finding of that study was the development of unsymmetrical, oscillating recirculation zones in the duct under some flow conditions. If $\dot{m}_j$ is the total jet mass-flow rate and Q is the total mass-flow rate in the duct and h and h_2 are the half-heights of the jet and the duct, the development of the flow was found to be governed by the parameters $Q/\dot{m}_j$ and h_2/h, as shown in Fig. 80.

Some of the results in Ref. 32 are of direct interest here even though the cases studied were wakes and deficit ($U_j < U_\infty$) jets. The transverse variation of static pressure often encountered in planar free mixing problems was men-

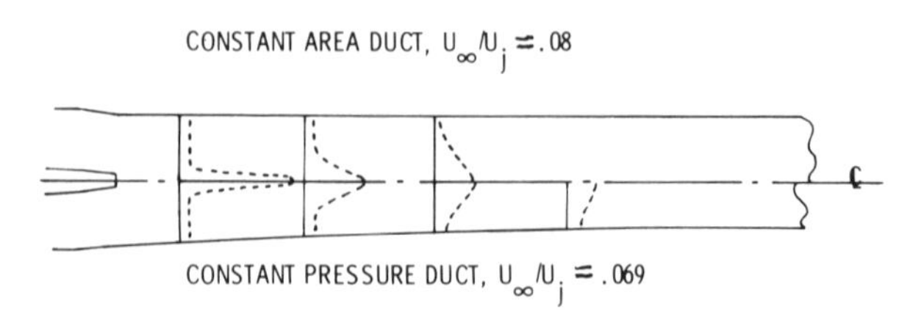

Fig. 76 Comparison of constant-area and constant-static-pressure axisymmetric jet mixing[106] (published with permission).

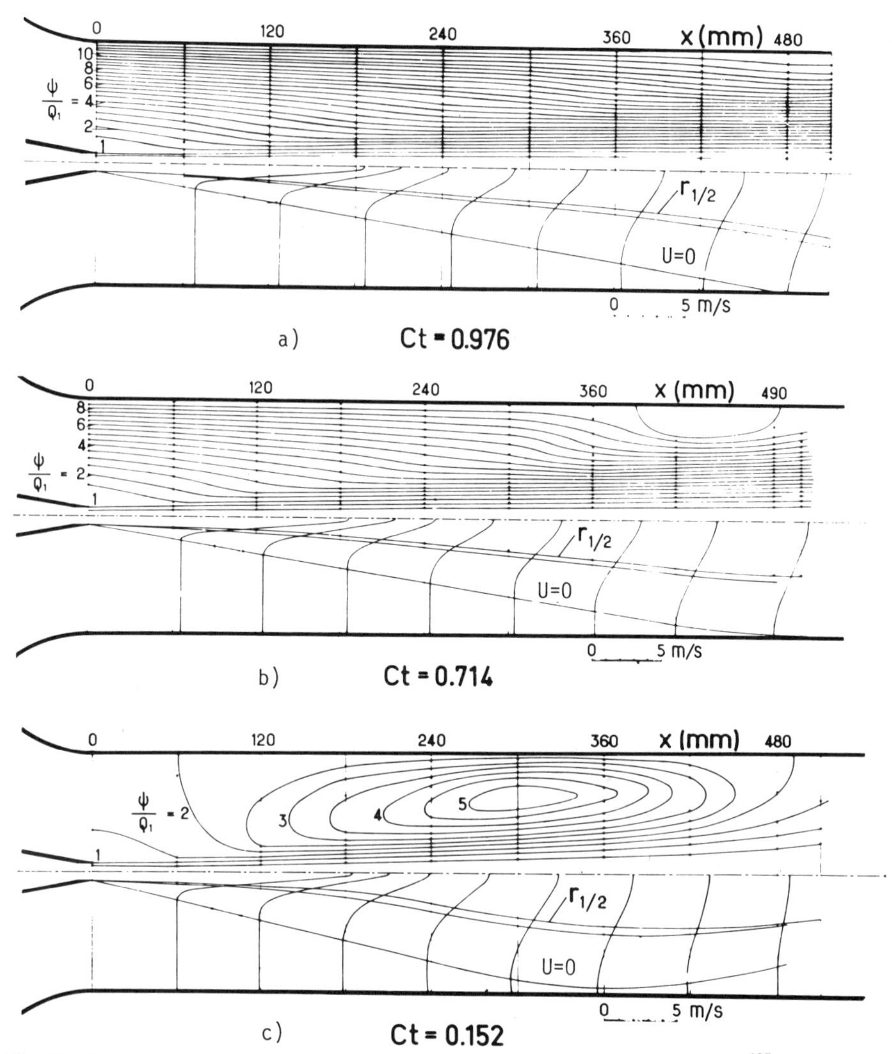

Fig. 77 Streamline patterns in a constant-area axisymmetric duct with a central jet[107] (published with permission).

tioned before. It is important to ask whether an imposed axial static pressure gradient influences the development of a transverse gradient. In Fig. 81, some results for the pressure difference across the viscous zone for a planar wake case are shown. Clearly, this large a difference cannot be neglected in analysis, and the boundary-layer form of the equations of motion can be expected to be deficient in such cases. The magnitude of the transverse static pressure difference was, however, found to be much smaller for axisymmetric and three-dimensional wakes.

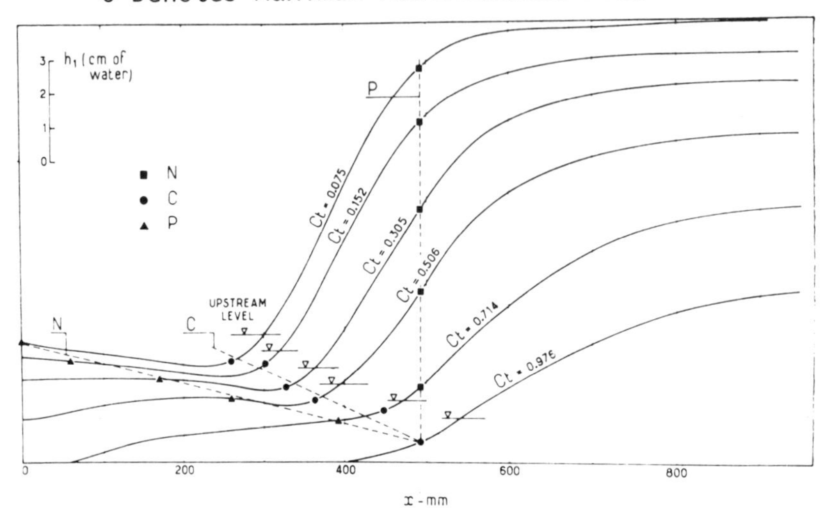

Fig. 78 Variation of wall static pressure in a constant-area axisymmetric duct with a central jet[107] (published with permission).

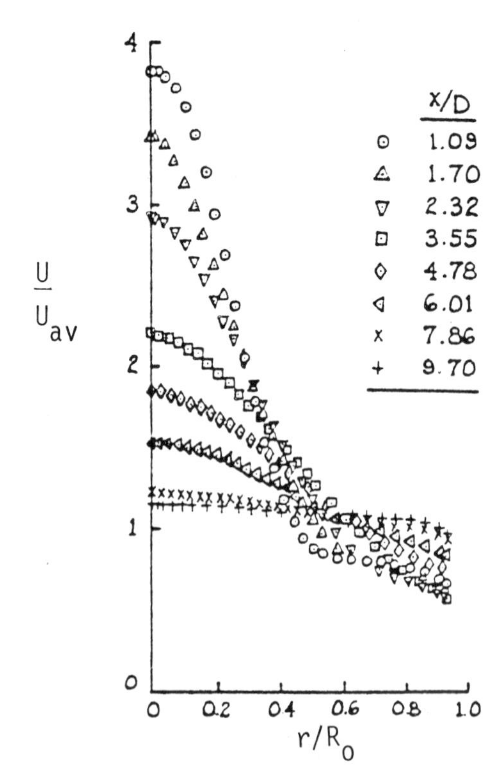

Fig. 79 Mean velocity profiles for constant-area ducted axisymmetric jet mixing with $U_\delta(0)/U_j = 0.25$[109] (published with permission).

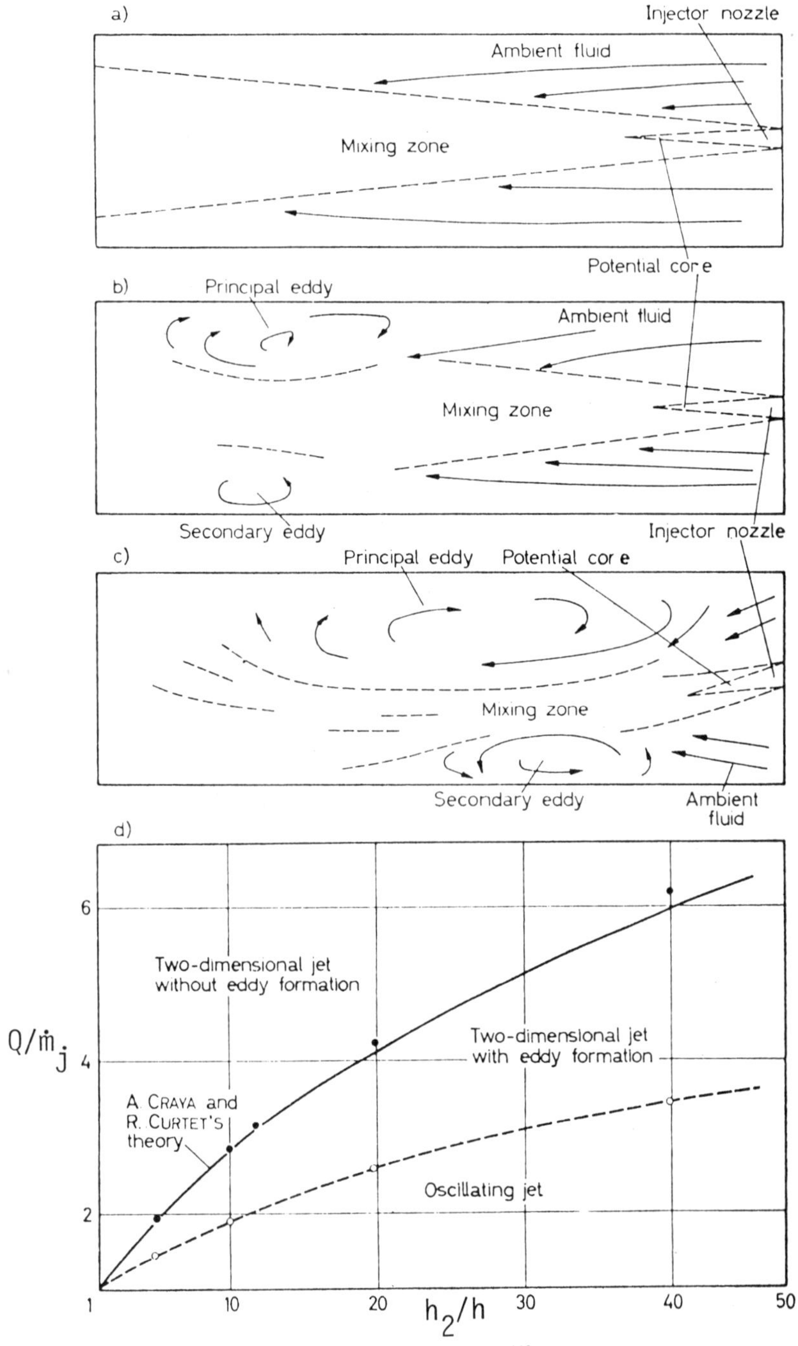

Fig. 80 Oscillating behavior of ducted planar jets[110] (published with permission).

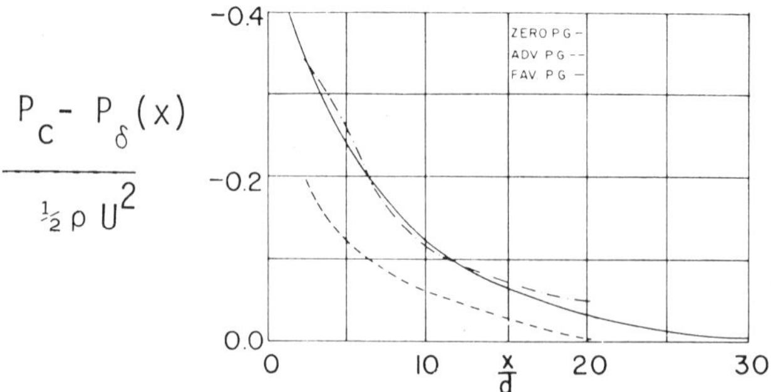

Fig. 81 Measured lateral static pressure difference across a planar wake in a pressure gradient.[32]

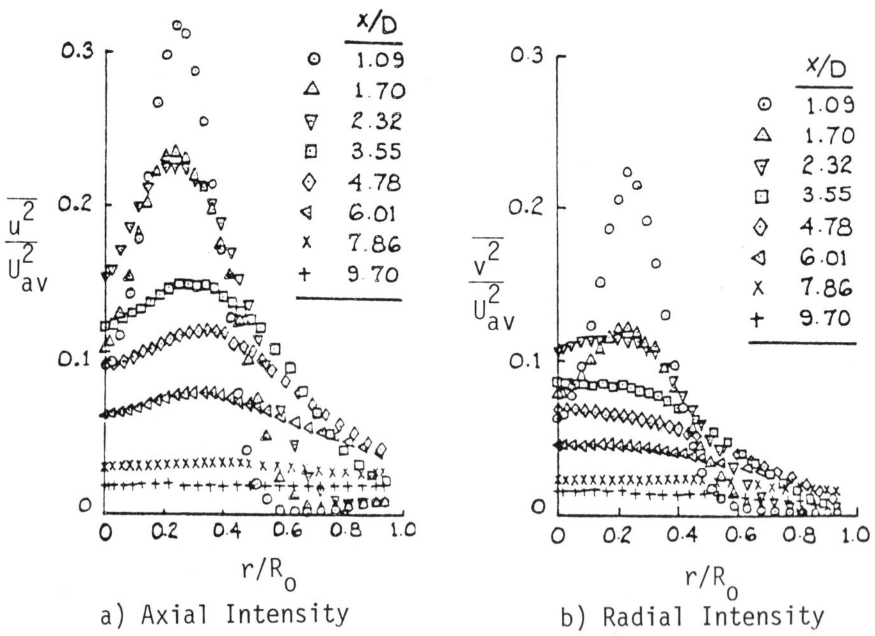

Fig. 82 Turbulence intensities for constant-area ducted axisymmetric jet mixing with $U_\delta(0)/U_j = 0.25$[109] (published with permission).

2. *Turbulence Data*

The turbulence data presented here are taken from Ref. 109. Figures 82a and 82b show the turbulence intensities in the axial and radial directions for the same conditions as Fig. 79. The maximum values of the axial intensities correspond roughly to those in an unconfined jet at similar conditions[27] when an attempt is made to account crudely for the axial variation of the "freestream" velocity in the duct problem. Reynolds stress data are shown in Fig. 83. It is interesting that the results given for large x/D do not yet correspond to those to be expected in a fully developed pipe flow. The mean velocity profiles are flatter near the axis, and the turbulent stresses are much higher.

It should not be surprising that the turbulence structure of the flow is increased markedly when recirculation exists. Figure 84 shows axial turbulence data for a case with large recirculation (see Fig. 77c). Beyond the center of the recirculation eddy, the intensity increases rapidly to truly large values.

The effects of both adverse and favorable pressure gradients on the quantity $\sqrt{\overline{u_c^2}}/\Delta U$ were studied for wakes and deficit jets in Ref. 32. The results show that a substantial influence exists; positive dP/dx increases this quantity.

B. Analysis

1. *Mean-Flow Models*

There are a number of approximate methods of analysis in the literature that employ either the integral type of analysis using profile similarity, e.g. Refs. 111-113, or a linearized approximation to the equations of motion.[114] Although these methods have achieved some success in predicting gross features of the flow development, they have now been superseded by direct numerical methods for predicting details of the flow.

We have already seen in Fig. 51 that the effects of a small pressure gradient are adequately accounted for by using a "constant-pressure" eddy viscosity model within a numerical solution of the exact (boundary-layer) equations of motion. For this case, transverse static pressure variations were not important.

Two recent studies[109,115] have compared the predictions obtained with numerical solutions and various turbulence models and data for confined,

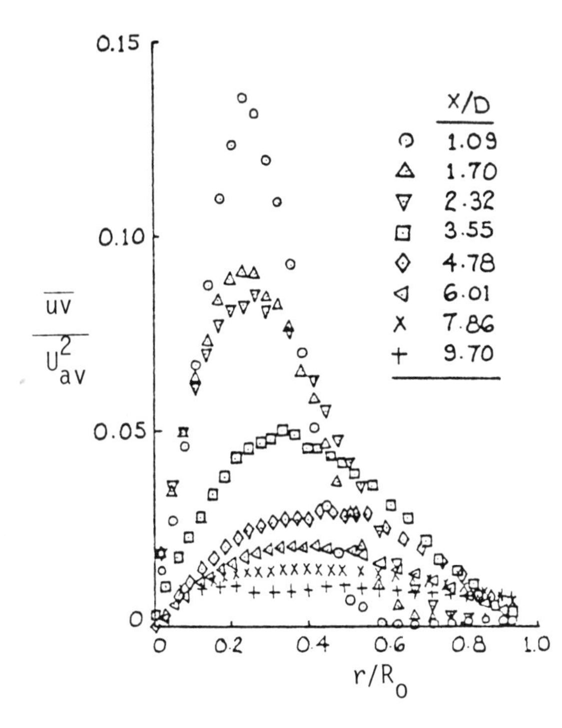

Fig. 83 Reynolds stress profiles for constant-area ducted axisymmetric jet mixing with $U_\delta(0)/U_j = 0.25$[109] (published with permission).

axisymmetric jets. Their results will comprise most of the remainder of this and the following subsections. Each paper employed its own experimental data, and the cases in Ref. 115 involved substantial recirculation, whereas those in Ref. 109 did not.

The simplest mean-flow model was the mixing length model used in Ref. 115:

$$l_m = 0.02(x) \tag{118}$$

The results are compared to experimental profiles at various x/D in Fig. 85, where it is seen that a relatively good prediction is obtained with this mixing length model.

The mixing length model used for the calculations in Ref. 109 is algebraic but much more elaborate. The scheme employed was originally presented in Ref. 116 and is shown here in Fig. 86. Region 1 is a boundary layer, and regions 2 and 4 are uniform velocity zones. Region 3 is a freejet flow, region 6 is a developed duct flow, and region 5 is the transition between the two. A comparison of prediction and experiment for the centerline velocity variation is shown in Fig. 87, where quite good agreement is apparent.

2. *Algebraic Turbulence Function Model*

In this class of model, the proportionality constant in an eddy viscosity model is taken as a function of $\sqrt{u_c^2}/\Delta U$. Calculations using this type of model for a free mixing flow in a pressure gradient were presented in Ref. 32 for the case of a planar wake. Since that problem had been found to involve non-negligible lateral static pressure variations, the boundary-layer equations were not adequate. A formulation including an approximate lateral momentum equation was therefore adopted. The value of the proportionality constant in

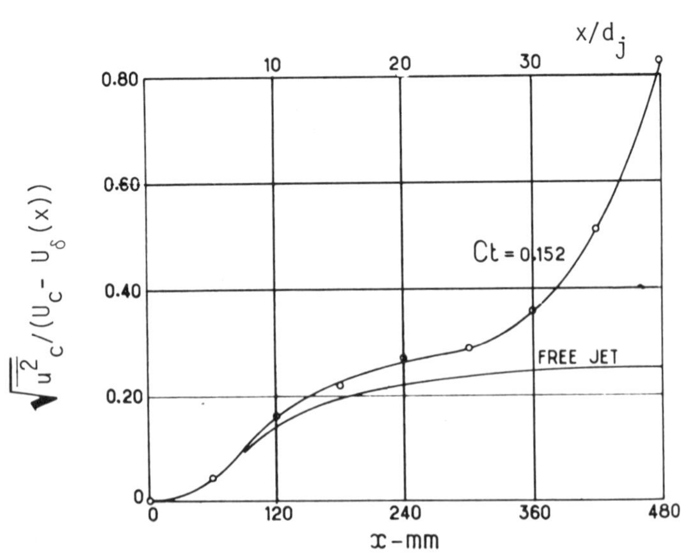

Fig. 84 Centerline turbulence intensity for constant-area ducted axisymmetric jet mixing[107] (published with permission).

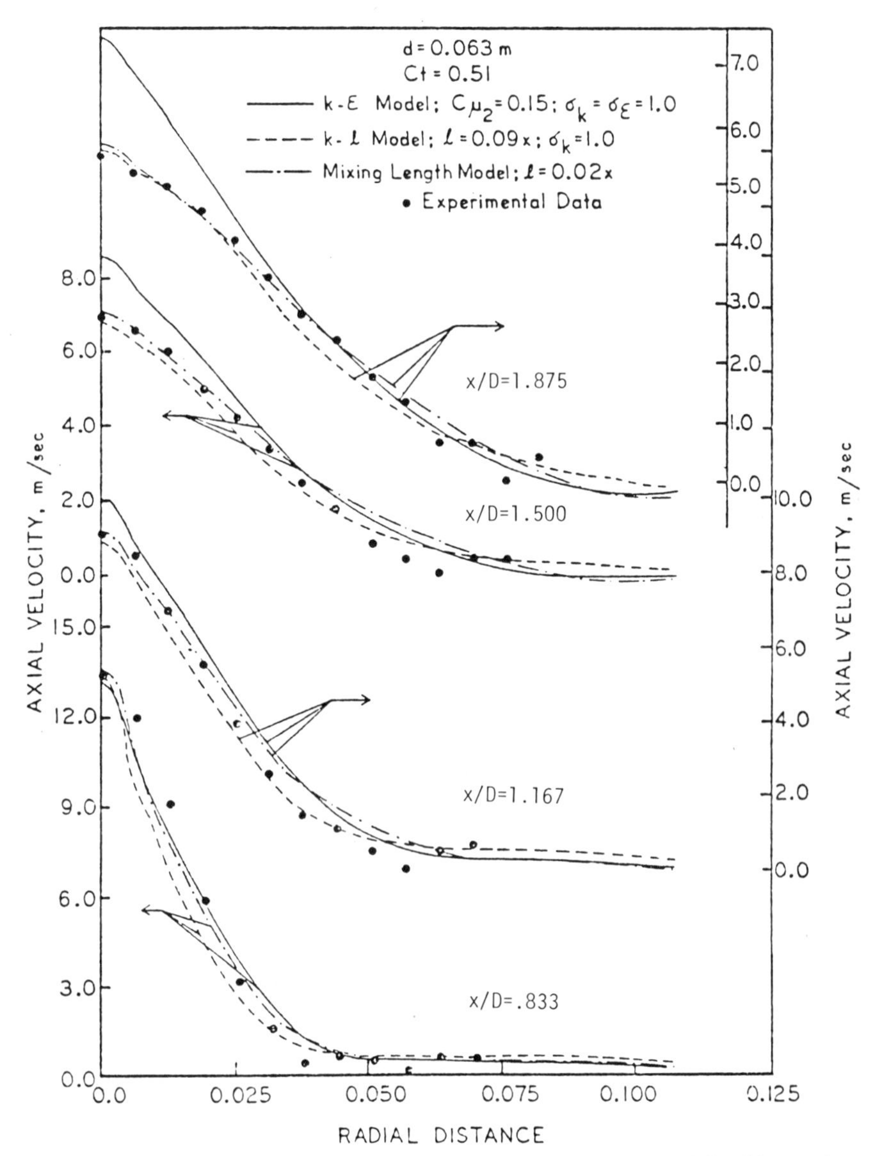

Fig. 85 Comparison of predictions with mean-flow and two-equation models with experiment for constant-area ducted axisymmetric jet mixing[115] (published with permission).

the eddy viscosity model was found by multiplying the value commonly used for jets by the ratio of the turbulence parameter $\sqrt{u_c^2}/\Delta U$ for the experimental case considered to that found in constant-pressure jets. The results of the calculation are compared to the data in Fig. 88 in terms of the centerline velocity variation and width. Good agreement was achieved. This was a very complex flow problem, since a mixed pressure gradient exists through the

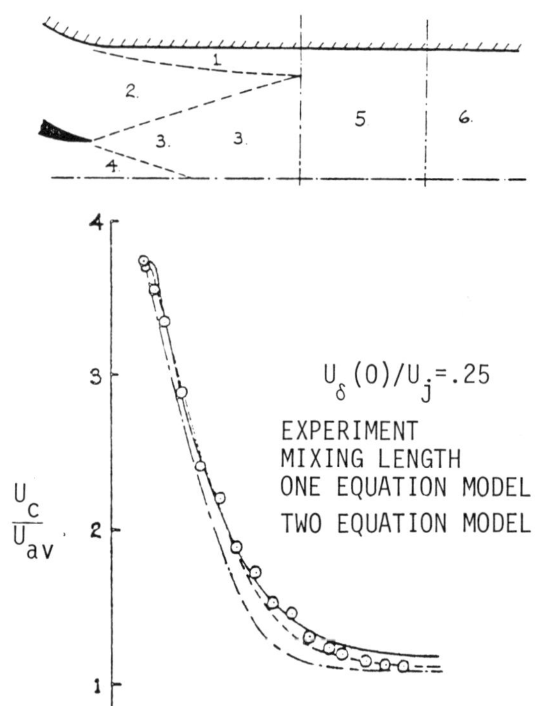

Fig. 86 Regions assumed in the mixing length model of Ref. 109 (published with permission).

Fig. 87 Comparison of predictions with mean-flow, one-equation, and two-equation models for constant-area ducted axisymmetric jet[109] (published with permission).

viscous zone. The pressure gradient along the centerline was adverse for the first 20 diam, whereas the external flow was favorable at all axial stations.

3. *One-Equation Models*

The one-equation model used for calculations in Ref. 115 is the same as that in Ref. 81. It is termed the *k-l* model in Fig. 85. There is not much to choose between the performance of this model and the simpler mixing length model.

The one-equation model employed in Ref. 109 uses essentially the same TKE equation, but the length scale l is determined from that discussed in Sec. III.B.1 based on Fig. 86 through the relation $l = l_m/1.83$. The predictions are denoted by the dashed line on Fig. 87. Again, the performance is comparable to that for the mixing length model.

4. *Two-Equation Models*

Reference 115 again follows the suggestions of Ref. 81, and relatively poor results are obtained, as may be seen in Fig. 85 for the solid curves

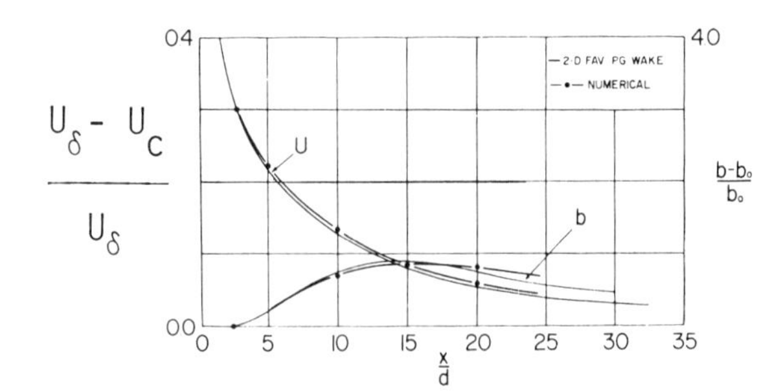

Fig. 88 Comparison of prediction with an algebraic turbulence function model and experiment for a planar wake in a favorable pressure gradient.[32]

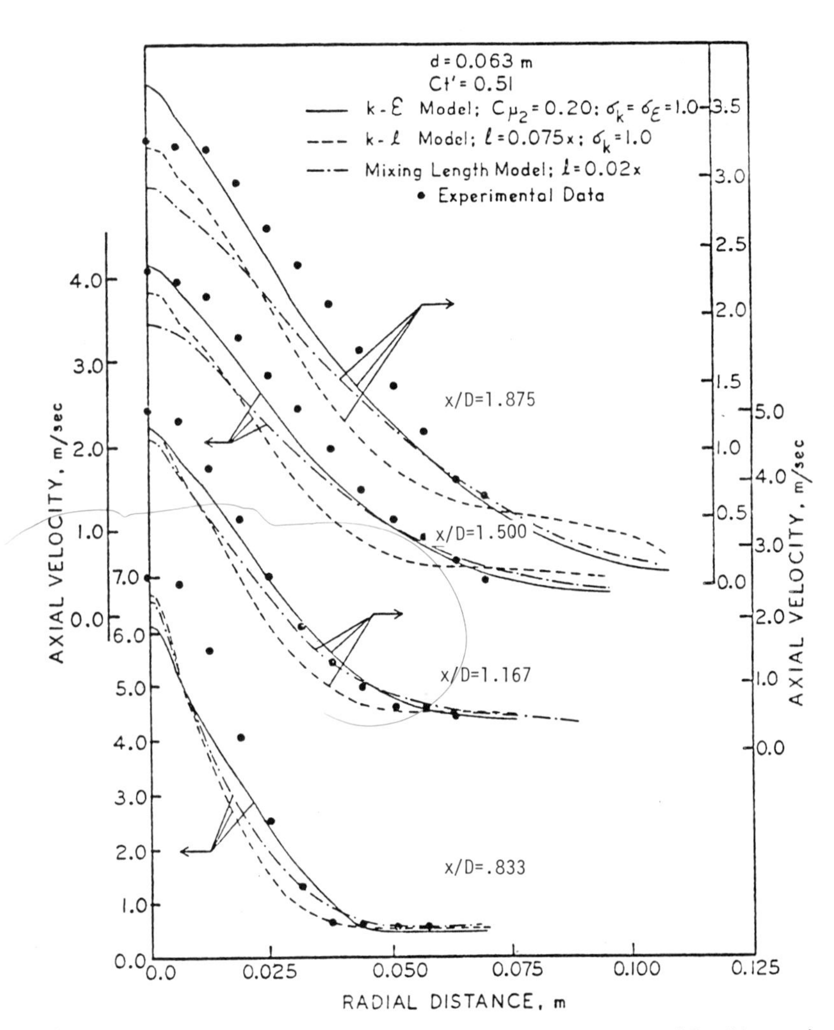

Fig. 89 Comparison of predictions with mean-flow and two-equation models with experiment for constant-area ducted axisymmetric heated jet mixing[115] (published with permission).

corresponding to what is called the "k-ϵ model." Experimental cases with a heated jet were also considered in this work, and there the two-equation model provided the best predictions. However, none of the predictions were very good, as shown in Fig. 89.

In Ref. 109, the two-equation model employed is also of the k-ϵ type, although the constants employed are similar but not the same as in Ref. 81. The comparisons in Fig. 87 again show this level of model having the poorest agreement with the data.

5. *Discussion*

There are no calculations available for problems with substantial pressure gradients based on either the Reynolds stress models or the direct turbulence models, and so comparisons of predictions and data cannot be presented. In leaving the subject of pressure gradient effects, it is interesting to note that any explicit effect of pressure gradient on the various models has been virtually ignored. The pressure gradient generally enters explicitly only through the dynamics of the flow as expressed by the equations of motion, not the turbulence models. The shape of the the profiles, for example, is affected by the pressure gradient, and this affects the actual values of the transport coefficients predicted by the models, but only implicitly. The pressure gradient influences the algebraic turbulence function model directly through the empirical values of $\sqrt{\overline{u^2_c}}/\Delta U$. The two equation model of Ref. 81 contains a weak dependence on pressure gradient through the term involving dU_c/dx in the proportionality constants [see Eqs. (85-87)], and this was included in the calculations of Ref. 115, but apparently without any salutory effect, at least for the isothermal cases.

IV. ZERO NET MOMENTUM DEFECT CASES

A. Background

In the general case, jet (and wake) mixing flows have a nonzero value of the net momentum flux in the viscous region [see Eq. (57)]; however, there are some cases where the integrated values of the momentum flux is zero. The wake behind a self-propelled body is the simplest flow that can be used to illustrate such a situation. This type of flow is of interest for two particular reasons. First, it has been found that the mean flow relaxes to nominally uniform conditions much more rapidly than the turbulent flow, so that some of the relationships between the mean and turbulent fields that have proven useful for other cases fail here. Second, in order to obtain a zero momentum flux integral, the velocity profiles almost by definition contain a number of inflection points. This makes the algebraic specification of length scales especially difficult.

B. Experimental Results

1. *Jet/Wake Combinations*

There have been two major experimental efforts that dealt with jet/wake combinations. At the University of Iowa,[117-119] the case of a circular disk with a concentric jet was studied. The flow from the jet was adjusted until its thrust exactly cancelled the drag of the disk. The resulting flowfield is shown schematically in Fig. 90.

Radial profiles of the mean flow are shown in Fig. 91, where the pressure P is referred to the pressure in the freestream and $r_{1/2}(x)$ is determined based on the turbulence intensity profile, not the mean velocity profile. Profiles of the turbulence intensity and the shear are given in Figs. 92 and 93. The axial variation of all of the measured flow quantities is shown in Table 8, where x_0 is the virtual origin shift for the flow assuming similarity of profiles. The value found of $x_0/D = 2$ implies that the virtual origin was downstream of the injection point for this case.

We have shown before that usual jets and wakes generally show a roughly constant value of the quantity $(\overline{u^2})_{max}/(\Delta U)^2$. This was also found to be an implicit assumption in eddy viscosity models (see Sec. II.C.3). Examining a plot of the data in Table 8 quickly shows that a relation of that type does not hold in the present case; ΔU decays much more rapidly with x than $\sqrt{\overline{u^2}}$. It can, therefore, be anticipated that conventional eddy viscosity models and perhaps other types of models developed for the more usual mixing cases will not be adequate here.

The experiment of Ref. 120 employed a thick ring to generate a wakelike flow with drag and a center jet for thrust. The mean-flow profiles are similar in form to those shown in Fig. 91. The axial variation of the mean centerline velocity and two turbulence intensities are shown in Fig. 94, along with corresponding results for a pure jet with the same value of U_j/U_∞. Here, one can see that the decay rate of not only the mean flow but also the turbulence intensities is accelerated for the zero net momentum flux case.

2. *Self-Propelled Bodies*

In this section, some results for wake development behind slender bodies with propulsors located at the stern are presented. Although this is not an "injection" type of flow, it bears a close resemblance to the flows in Sec. IV.B.1, and much of the available analytical work for flows with a zero net momentum flux has been directed at these cases.

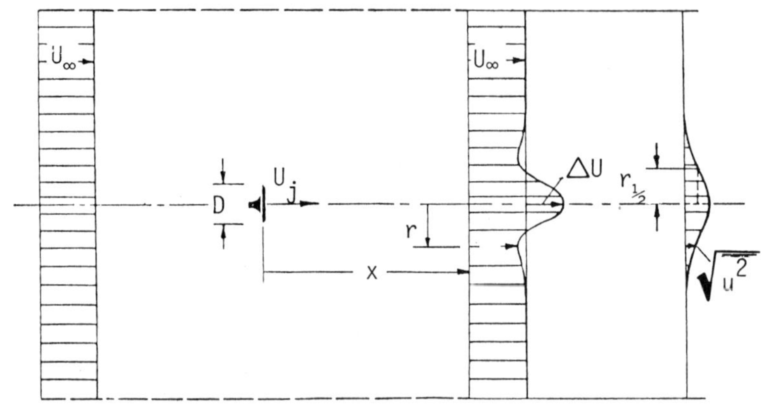

Fig. 90 Schematic of the flow studied in Ref. 119 (published with permission).

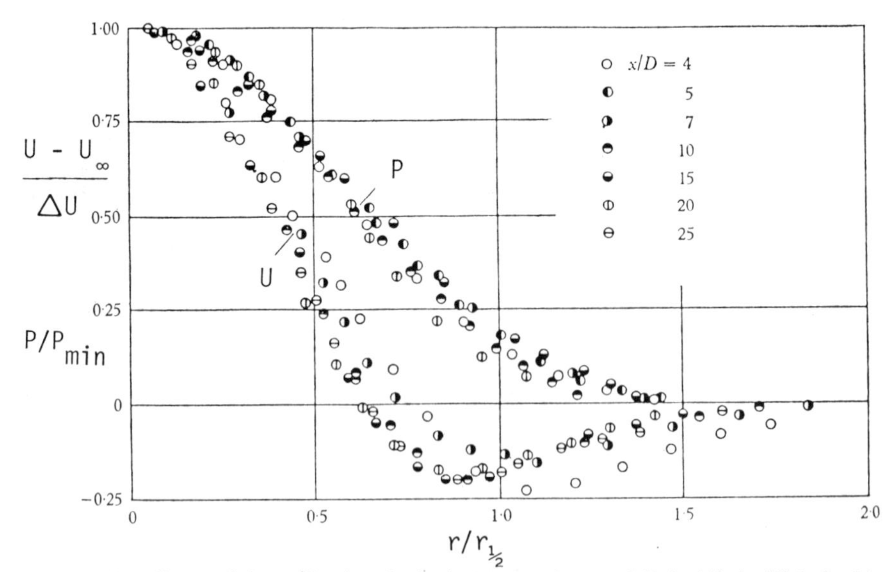

Fig. 91 Mean-flow radial profiles for the jet/wake experiment of Ref. 119 (published with permission).

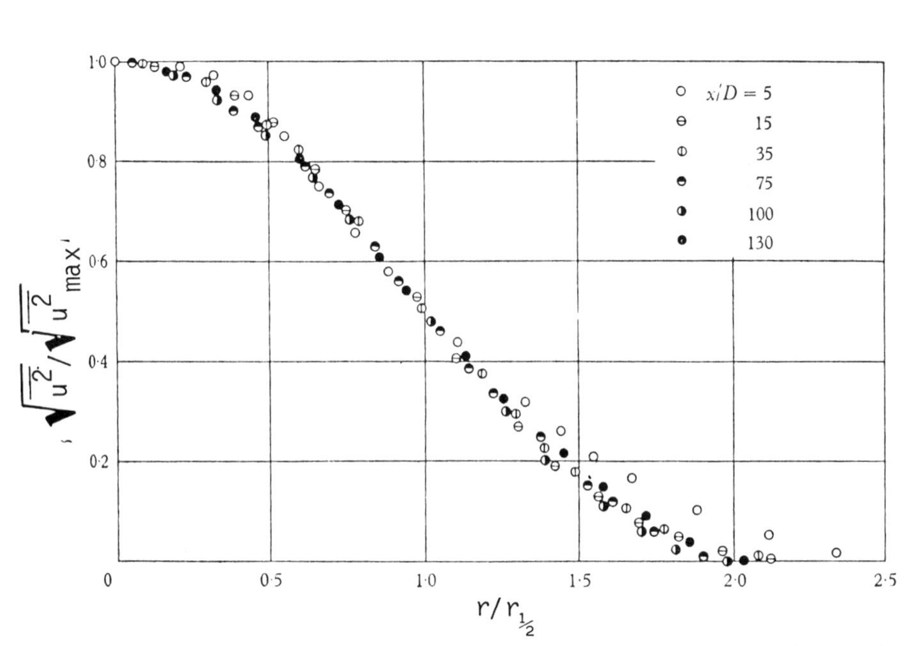

Fig. 92 Radial profile of axial turbulence intensity for the jet/wake experiment of Ref. 119 (published with permission).

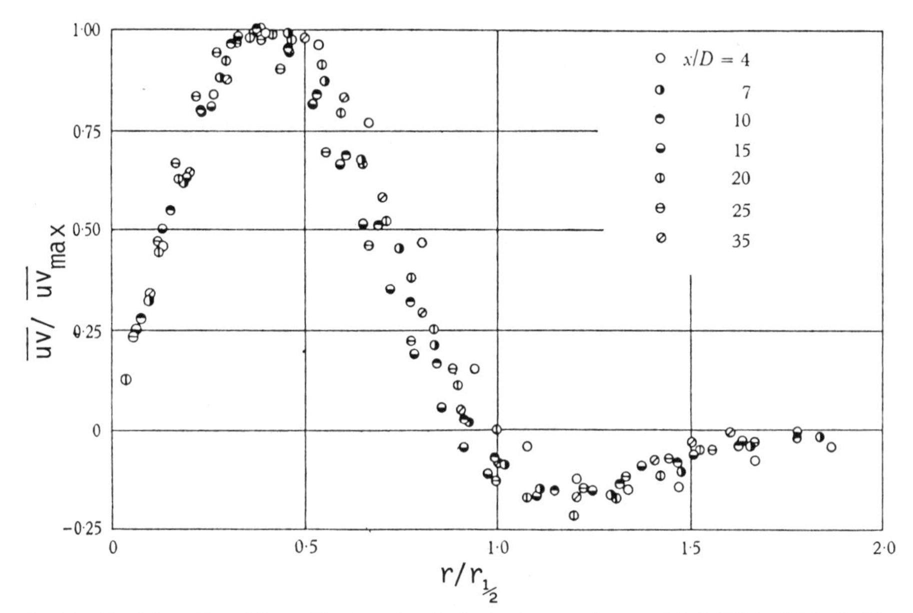

Fig. 93 Radial profile of Reynolds stress for the jet/wake experiment of Ref. 119 (published with permission).

Table 8 Axial variation of flow variables for jet disk flow

x/D	$(x-x_0)/D$	$10^2 P_{min}/\frac{1}{2}\rho U_\infty^2$	$\Delta U_{max}/U_\infty$	$10^2(\overline{uv}_{max}/U_\infty^2)$	$\sqrt{\overline{u^2}}_{max}/U_\infty$	$\sqrt{\overline{v^2}}_{max}/U_\infty$	$\sqrt{\overline{w^2}}_{max}/U_\infty$	$10^2(k/U_\infty^2)$	$r_{1/2}/D$
4	2	−7.36	0.363	1.620	0.246	0.198	0.193	13.70	0.515
5	3	−4.10	0.252	0.656	0.163	0.127	0.124	5.808	0.60
7	5	−2.10	0.134	0.307	0.110	0.0993	0.0988	3.172	0.72
10	8	−1.05	0.059	0.131	0.0745	0.0700	0.0660	1.481	0.875
15	13	−0.385	0.0277	0.051	0.0535	0.0452	0.0446	0.689	1.025
20	18	−0.27	0.0149	0.024	0.0425	0.0355	0.0335	0.418	1.12
25	23	...	0.0092	0.014	0.0330	0.0310	0.0280	0.283	1.21
35	33	...	...	0.0065	0.0239	0.0232	0.0217	0.158	1.35
50	48	...	...	...	0.0170	0.0162	0.0151	0.078	1.52

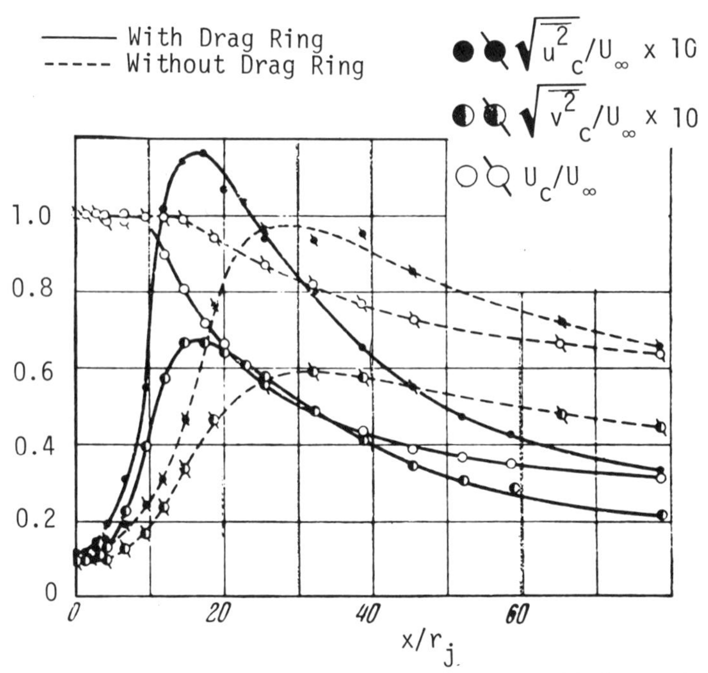

Fig. 94 Streamwise variation of centerline velocity and turbulence intensities for an axisymmetric jet with and without a concentric drag ring to balance thrust and drag[120] (published with permission).

The data of Ref. 121 have been selected for presentation here, since they provide a direct comparison of three cases, all with the same forebody shape: 1) pure drag body, 2) self-propelled by axial fluid injection through a peripheral slot with a centerbody, and 3) self-propelled by mounting a rotating propeller on the drag body. The development of the mean flow in the three cases is shown in Figs. 95a-95c. The more complicated nature of the flow for the self-propelled cases is clear.

The axial variation of the maximum values of the axial turbulence intensity and the shear is presented in Figs. 96 and 97. For the propeller-driven model, the axial turbulence intensity was actually less than that in the two transverse directions, whereas for the drag body and the jet-propelled body, the axial turbulence was the largest, as usual.

C. Analysis

1. Mean-Flow Models

In discussing the experimental results, we noted that conventional eddy viscosity models could be expected to have difficulty with these flows. Indeed, this is the case. The use of the successful jet model of the Ref. 77 for the case of Refs. 117-119 underpredicted the axial velocity decay rate by a large factor.[122] It was possible to construct a new model that serves well for jet/wake

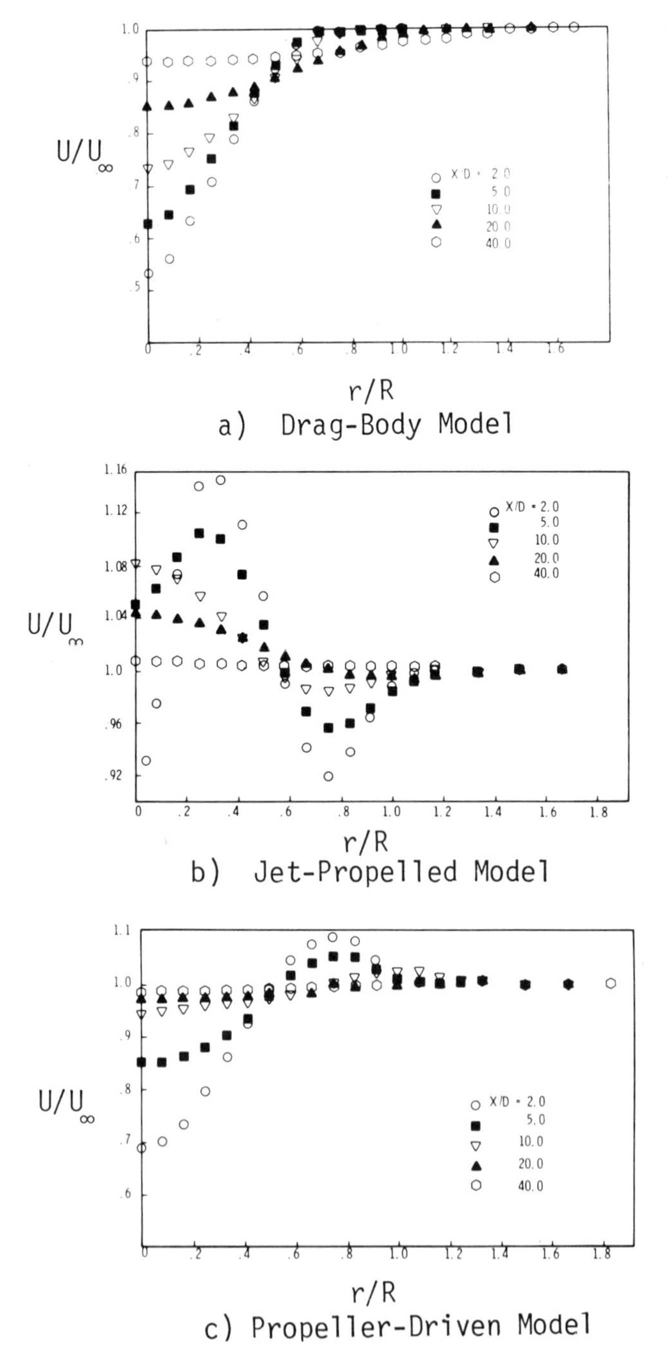

Fig. 95 Mean-flow profiles in the wake behind a self-propelled body.[121]

combinations,[122] but the model has not proven general enough for use for the wakes behind self-propelled bodies.

2. *Algebraic Turbulence Function Model*

Since the basic premise of this approach was the constancy of $\overline{u_m^2}/(\Delta U)^2$ and this is not supported by the data in the present case, this type of model is inappropriate here.

3. *One-Equation Models*

Reasonable success has been obtained with this level of turbulence model. The method of Harsha[82] was applied in a slightly modified form to the cases of Ref. 121 in Ref. 123. Recall that Ref. 82 utilizes the Bradshaw relation for shear and turbulent kinetic energy:

$$\tau_T = a_l \rho k \tag{119}$$

where $a_l = 0.3$. Although a_l can be taken at a constant value of 0.3 over most of the wake flow, it is not a constant in the vicinity of the centerline, since the shear stress vanishes at that point for axisymmetric flows; in Ref. 123 the following was used:

$$a_l = 0.3 \frac{\partial U/\partial r}{|\partial U/\partial r|_{\max}} \tag{120}$$

for normal shear flows, i.e., drag body case 1. This relation was applied from the centerline to the point $\partial U/\partial r = (\partial U/\partial r)_{\max}$. Beyond that point

$$a_l = 0.3 \frac{\partial U/\partial r}{|\partial U/\partial r|} \tag{121}$$

was used which yields the proper sign. To avoid an excessive shear stress value in the outer region of the flow for cases with a zero net momentum defect, the

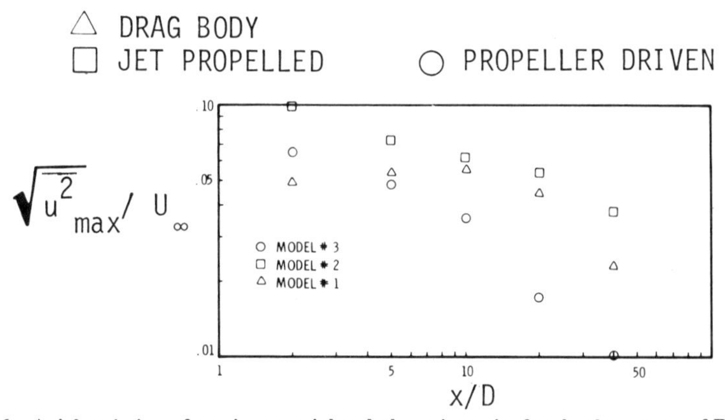

Fig. 96 Axial variation of maximum axial turbulence intensity for the three cases of Ref. 121.

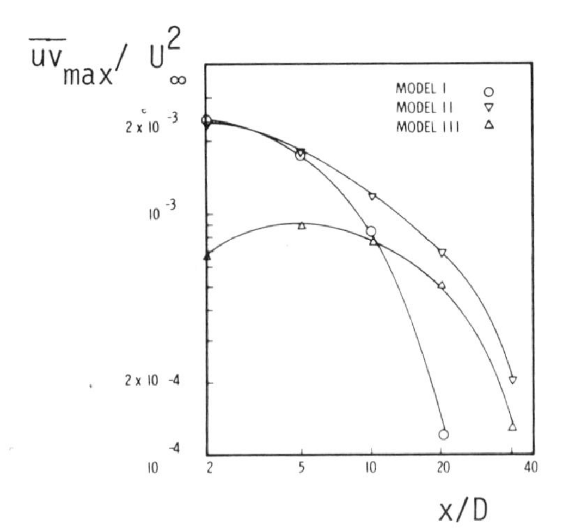

Fig. 97 Axial variation of maximum Reynolds stress for the three cases of Ref. 121.

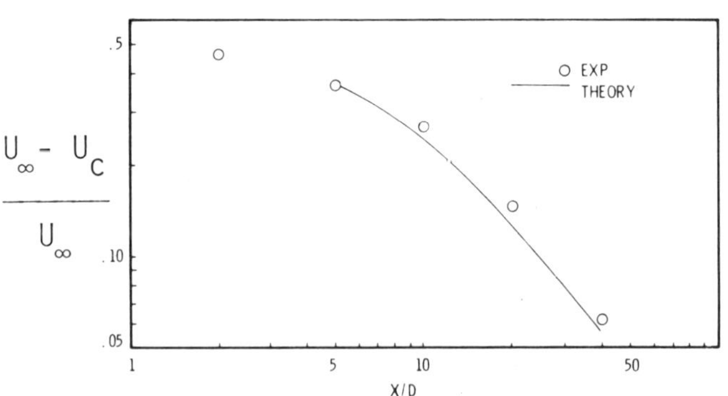

Fig. 98 Comparison of a prediction with a one-equation model and experiment for the wake of a slender drag body.[123]

program was modified to include the following formulation:

$$a_l = 0.3 \frac{\partial U/\partial r}{|\partial U/\partial r|_{\max}}; \qquad \frac{\partial U}{\partial r} > 0 \tag{122a}$$

$$a_l = 0.3 \frac{\partial U/\partial r}{|\partial U/\partial r_{\min}|}; \qquad \frac{\partial U}{\partial r} < 0 \tag{122b}$$

These modifications were suggested by Harsha in a private communication in 1974.

Note that the use of Eq. (119) reduces the importance of the length scale formula chosen, since l now plays a less influential role than if Eq. (12) with Eq. (18) were used.

Since the formulation is parabolic, the principal information required to undertake a calculation is in the form of initial conditions. These were taken from the experimental measurements at $x/D=5$. This included not only the radial distributions of the dependent variables, but also an initial eddy viscosity distribution which was obtained from the measured shear stress and velocity distributions. For case 2, this procedure produced results in strong disagreement with the data, so those results are not presented here. The reasons for the poor performance of the analysis for jet-propelled case 2 have never been clear.

The routine of Ref. 123 did not include a tangential momentum equation, so that swirl is not accounted for, and the calculations for propeller-driven case 3 are not exact and must be interpreted carefully. The principal calculations were begun at $x/D=5$, since the swirl velocity had decayed to a low value at that station.

For drag-body case 1, the numerical prediction of the downstream decay of the centerline mean velocity deficit, as shown in Fig. 98, agrees well with the experimental results. As shown in Fig. 99, the numerical computations for case 3 overpredict the experimental values initially, and the rate of decrease of velocity deficit is considerably greater than that suggested by the data when $x/D=40$ is reached.

The TKE program also provided the streamwise variation in radial shear stress for each of the models. The computations for case 3 showed fairly good agreement with experiment; however, the rate of decay is lower than it should be.

The apparent success of this level of model for self-propelled cases prompted the use of an integrated TKE model [see Eq. (80a) and the accompanying discussion] within a Navier-Stokes equation treatment of the stern and propeller region in Refs. 124 and 125. However, in that case, swirl was treated directly, and we shall save a discussion of that material until the next chapter, which deals with that subject in detail.

4. *Two-Equation Models*

There are no calculations available in the literature for experimental cases with a zero net momentum flux, so that no assessment of this level of model

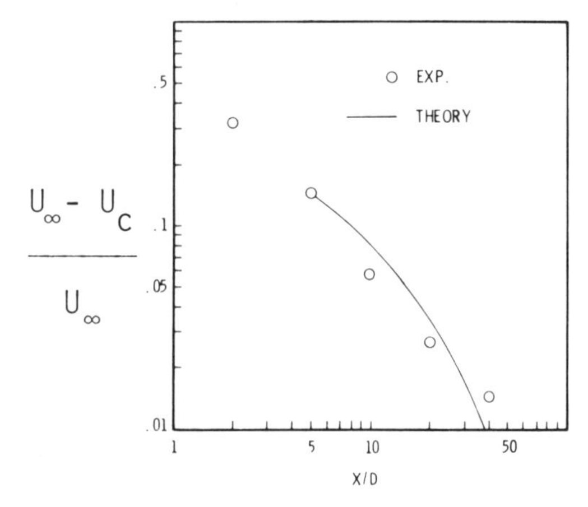

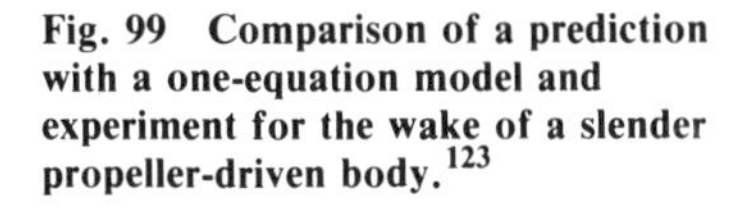
Fig. 99 Comparison of a prediction with a one-equation model and experiment for the wake of a slender propeller-driven body.[123]

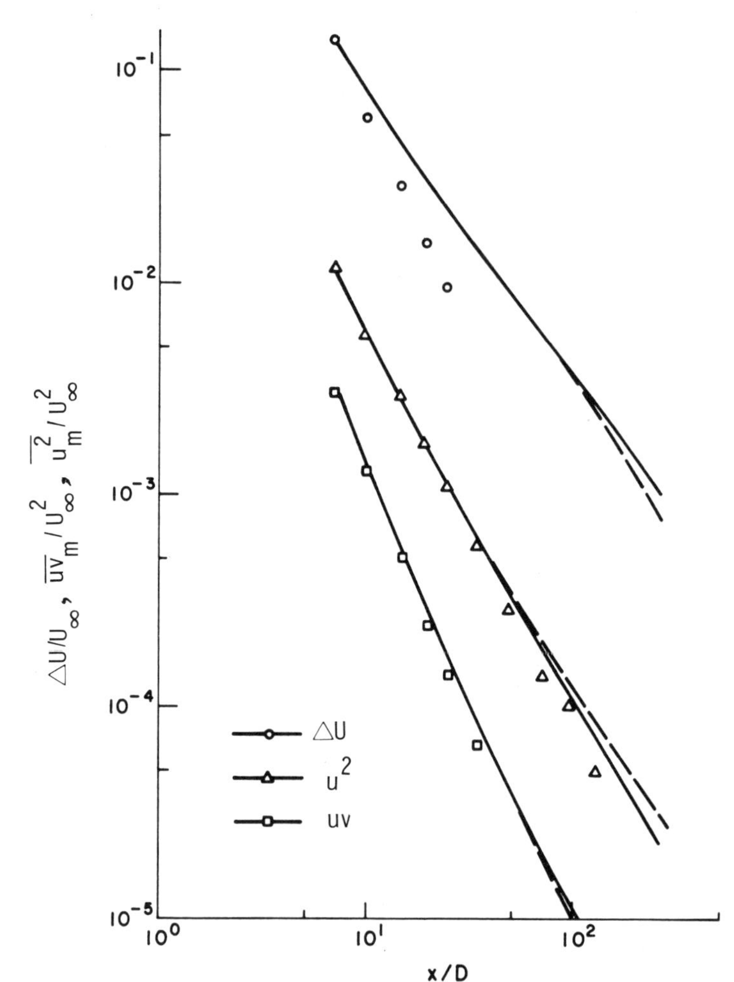

Fig. 100 Comparison of a prediction with a Reynolds stress model and the experiment of Ref. 119 for a jet/disk combination[126] (published with permission).

for such cases can be made. It might, however, be expected that the confusion surrounding the specification of an algebraic length scale relation for the complicated mean velocity profiles encountered in these cases might make this a good field for application of the two-equation models. They, of course, do not require a simple algebraic formula to be specified a priori.

5. *Reynolds Stress Models*

The Reynolds stress model of Donaldson has been applied to several cases of interest here in Ref. 126. Comparison with the jet/wake case of Refs. 117-119 is shown in Fig. 100. The prediction of the two turbulence quantities is actually better than that for the mean velocity.

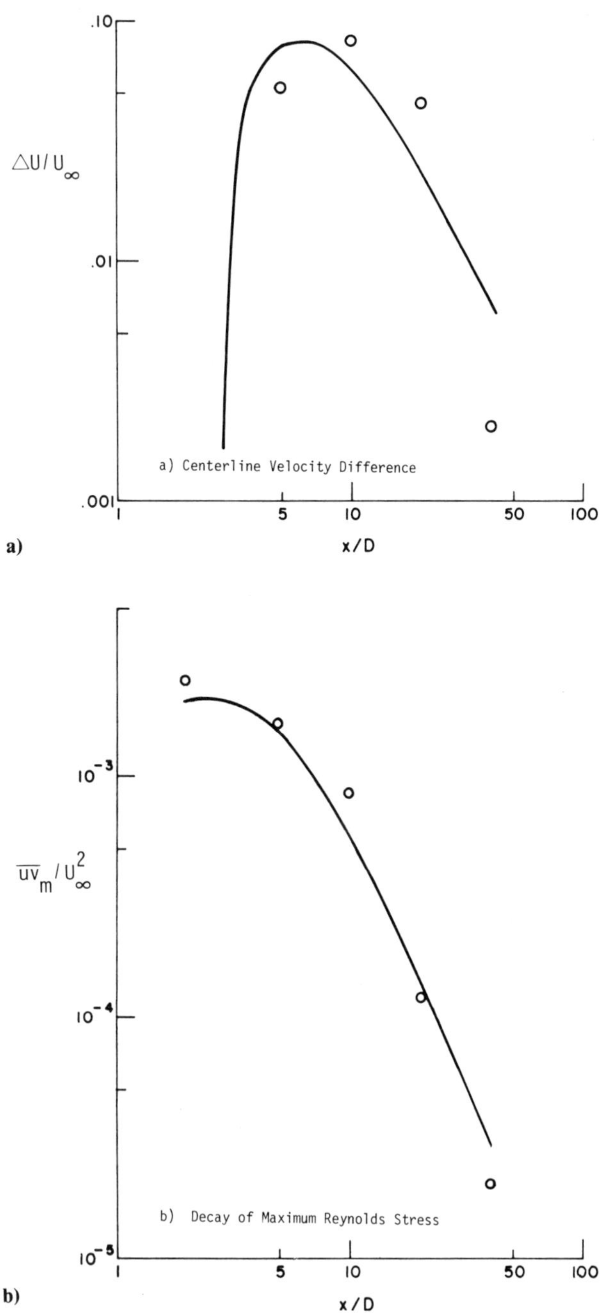

Fig. 101 Comparison of a prediction with a Reynolds stress model and the experiments of Ref. 121 for the wake behind a jet-propelled slender body[126] (published with permission).

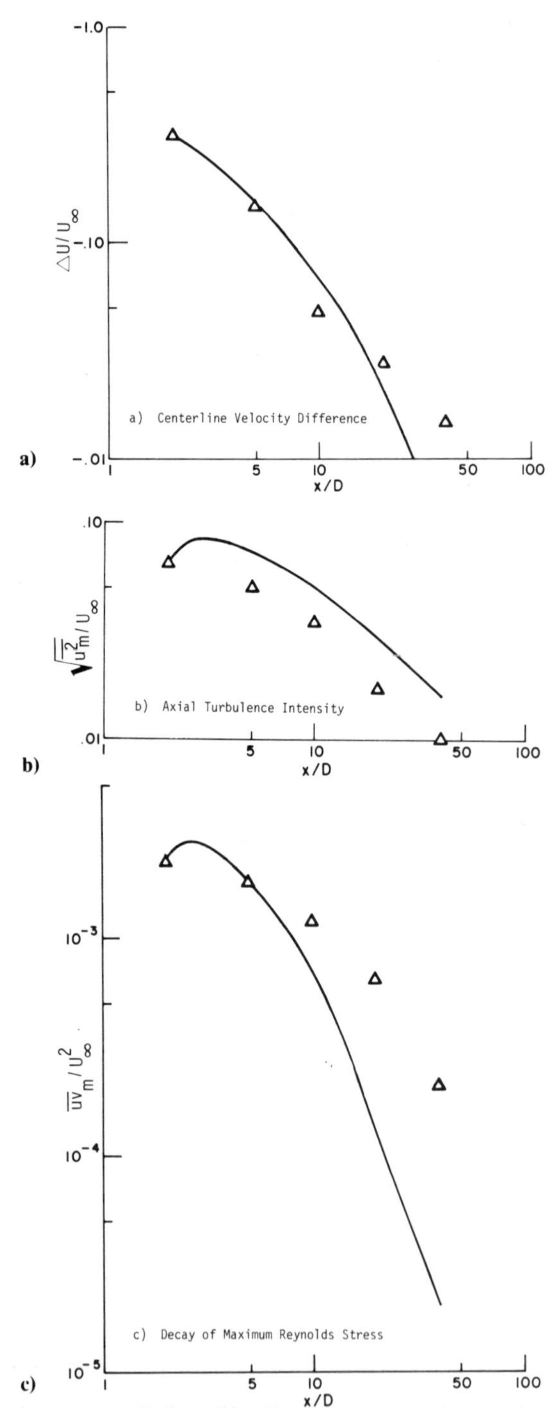

Fig. 102 Comparison of a prediction with a Reynolds stress model and the experiments of Ref. 121 for the wake behind a slender propeller-driven body[126] (published with permission).

Predictions for the two self-propelled (jet-propelled and propeller-driven) cases of Ref. 121 are shown in Figs. 101 and 102. The jet-propelled case is shown in Figs. 101a and 101b, where it can be seen that again the turbulence field is predicted more accurately than the mean field. The propeller-driven case, of course, involves substantial swirl and a quite different turbulent field than normal jets and wakes. The results in Figs. 102a-102c show that the predictions for the turbulence quantities are degraded somewhat, but that the agreement with the mean flow data is improved.

6. Discussion

No direct turbulence models have been applied to the present type of flow situation, and so no judgement as to their adequacy is possible. On the basis of the available results, it may be concluded that models of at least the one-equation level are absolutely required for these flows. The performance of the one Reynolds stress model for which results are available was impressive.

V. FLOWS WITH SWIRL

A. Background

There are a number of situations of practical interest where the analysis of mixing flows with swirl is important. In some cases, the swirl is intentionally introduced in an attempt to enhance mixing. In other cases, the swirl is a byproduct of another operation such as behind a propeller for propulsion or a windmill. This chapter will be restricted to flows where the swirl is initially present at some axial station and then freely relaxes as the flow proceeds downstream. Initial swirl can be easily produced by a propeller or rotating mixer of some sort or in a jet by vanes or peripheral fluid injection in the injector tube.

B. Experimental Information

1. Swirling Jets

The effects of various initial swirl levels on the development of the mean flow are shown in Figs. 103-105 taken from Ref. 18, which also contains data from some other work in the Soviet literature. Clearly, increasing the initial swirl level increases the rate of mixing of all of the mean-flow variables. Transverse axial velocity profiles are shown in Fig. 106, and transverse tangential (swirl) velocity profiles are shown in Fig. 107, both for the highest initial swirl level from the tests in Figs. 103-105. The outer portions of the tangential velocity profiles decay in a manner similar to an inviscid vortex, but the inner portion is dominated by viscous effects. The tangential velocity at the centerline is zero rather than the infinite value for an inviscid vortex.

A series of experiments for homogeneous flows with $U_\infty = 0$ has appeared.[127-129] In the English language literature, it is common to specify overall initial swirl rates by $S \equiv T^*/\mathfrak{M} r_j$, where $\mathfrak{M}$ is the integral of the pressure plus the axial momentum [Eq. (116)], and T^* is the total initial angular momentum:

$$T^* \equiv \int_A \rho U W \mathrm{d}A \tag{123}$$

The response of the width growth to variations in S is shown in Fig. 108. Axial velocity profiles for several values of S are shown in Fig. 109. For $S = 0.6$ and greater, there is substantial reverse flow on the axis. This is primarily a result of the large radial pressure gradient that is produced by the swirl. Measurements of radial static pressure profiles are given in Fig. 110. This radial pressure gradient is easily explained by reference to the momentum equation in the tangential direction, which in the boundary-layer form may be written as

$$-\frac{W^2}{r} = -\frac{1}{\rho}\frac{\partial P}{\partial r} - \frac{\partial \overline{v^2}}{\partial r} - \frac{\overline{v^2}}{r} + \frac{\overline{w^2}}{r} \tag{124}$$

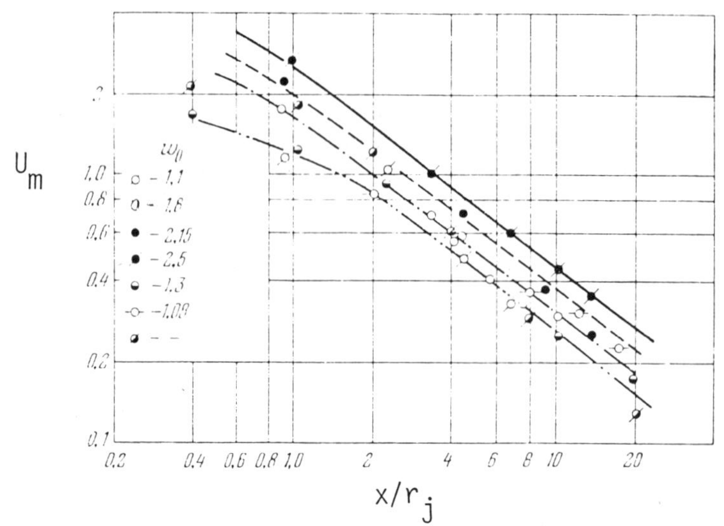

Fig. 103 Variation of maximum axial velocity as a function of initial swirl level and streamwise distance[17] (published with permission).

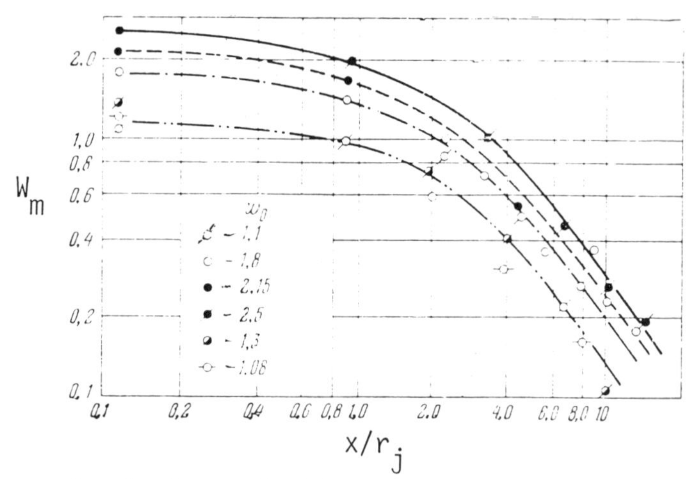

Fig. 104 Variation of maximum swirl velocity as a function of initial swirl level and streamwise distance[17] (published with permission).

The mean-flow terms show a simple, direct relationship between swirl magnitude and radial pressure gradient. When the outer fluid is stationary or, at least, not swirling, then the static pressure drops rapidly from the external value to a low value on the axis. The actual total variation is a function of swirl rate and is thus largest for small x/d where the swirl has not yet decayed significantly. The result is an adverse pressure gradient along the axis with increasing x/d. For large enough swirl, this gradient becomes strong enough to

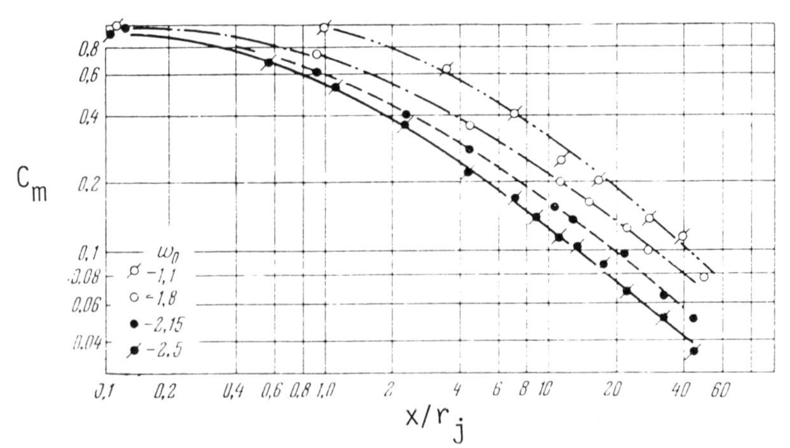

Fig. 105 Variation of the maximum value of concentration for foreign gas injection as a function of initial swirl level and streamwise distance[17] (published with permission).

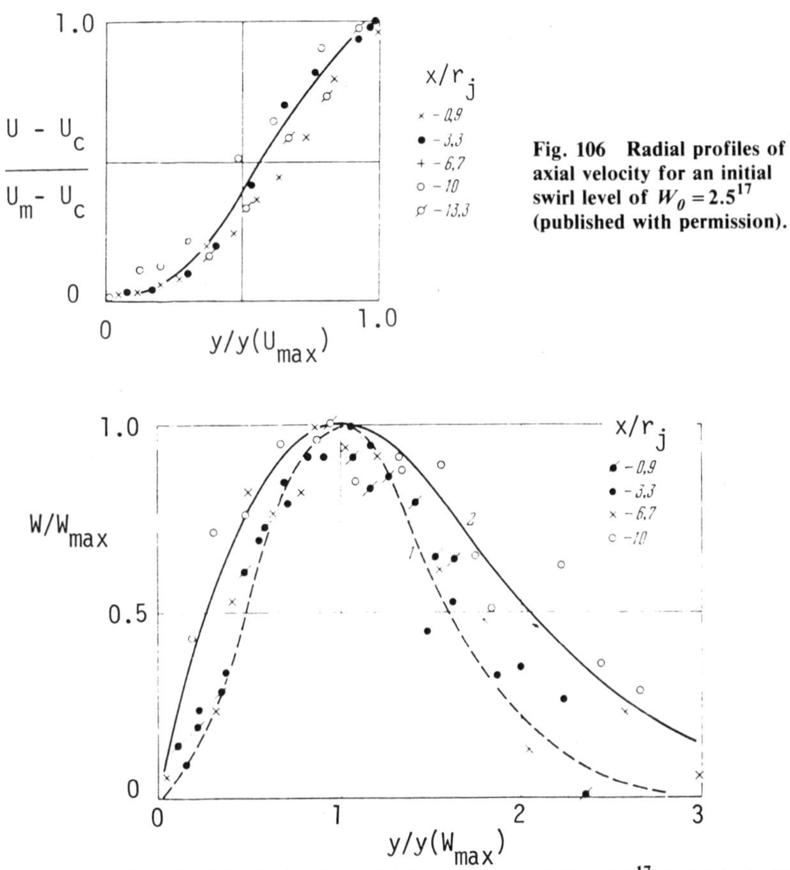

Fig. 106 Radial profiles of axial velocity for an initial swirl level of $W_0 = 2.5$[17] (published with permission).

Fig. 107 Radial profiles of swirl velocity for an initial swirl level of $W_0 = 2.5$[17] (published with permission).

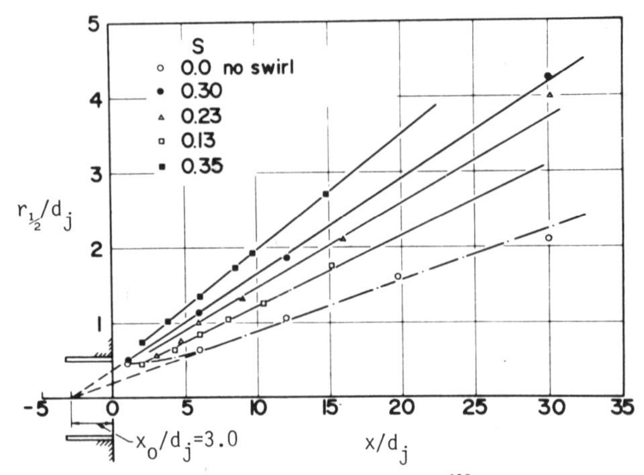

Fig. 108 Half-radius growth as a function of swirl number[129] (published with permission).

produce flow separation and reversal. In the Soviet literature, initial swirl level is denoted by $\Phi \equiv W_{max}/U_{max}$ at injection. In these terms, the limit for the occurrence of recirculation on the axis is reported in Ref. 18 as $\Phi = 0.6$. The length of the recirculation zone $l°$ can be described roughly by the relationship

$$l° \approx W_0 \tag{125}$$

where $W_0 = W_{max}/U_{av}$.

Other experimental studies for swirling jets with $U_\infty = 0$ may be found in Refs. 104 and 130, and we shall be referring to those data below. One of the few experiments on swirling jets with a high-velocity external stream is presented in Ref. 131, where the outer flow was supersonic ($M_\infty = 4.0$). Comparing the results of tests with and without initial swirl in the jet showed that swirl was not effective in promoting mixing with such a high-velocity external stream.

For turbulence data in swirling jets, use will be made here of the measurements in Ref. 104. However, since that information will play a critical role in the assessment of Reynolds stress models for use in swirling flows, we shall hold the presentation of the data until that section where predictions and data can be conveniently shown together.

2. *Wakes Behind Propeller-Driven Bodies*

Some information for such flows was presented in Figs. 95c, 96 and 97. The interested reader can find further details in Ref. 121 and the results of additional studies in Refs. 132 and 133. The experiments in Ref. 133 trace the behavior of a slight temperature perturbation in the approach flow over the body and through the propeller into the wake.

C. Analysis

1. *Mean-Flow Models*

One of the earliest treatments of swirling jets is in Ref. 134. In the far field, the equations of motion are linearized, and the Prandtl eddy viscosity model,

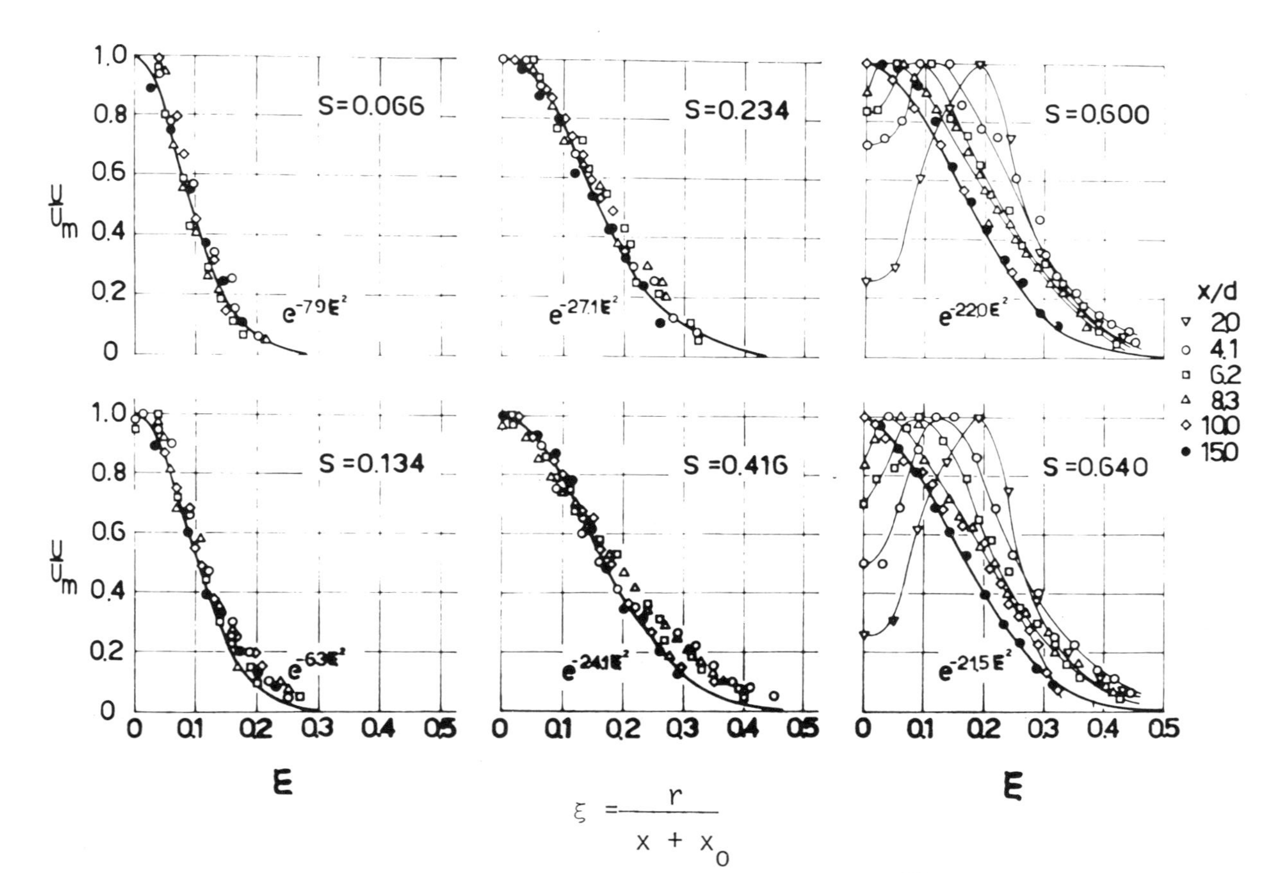

Fig. 109 Radial profiles of axial velocity in swirling jets as a function of swirl number[128] (published with permission).

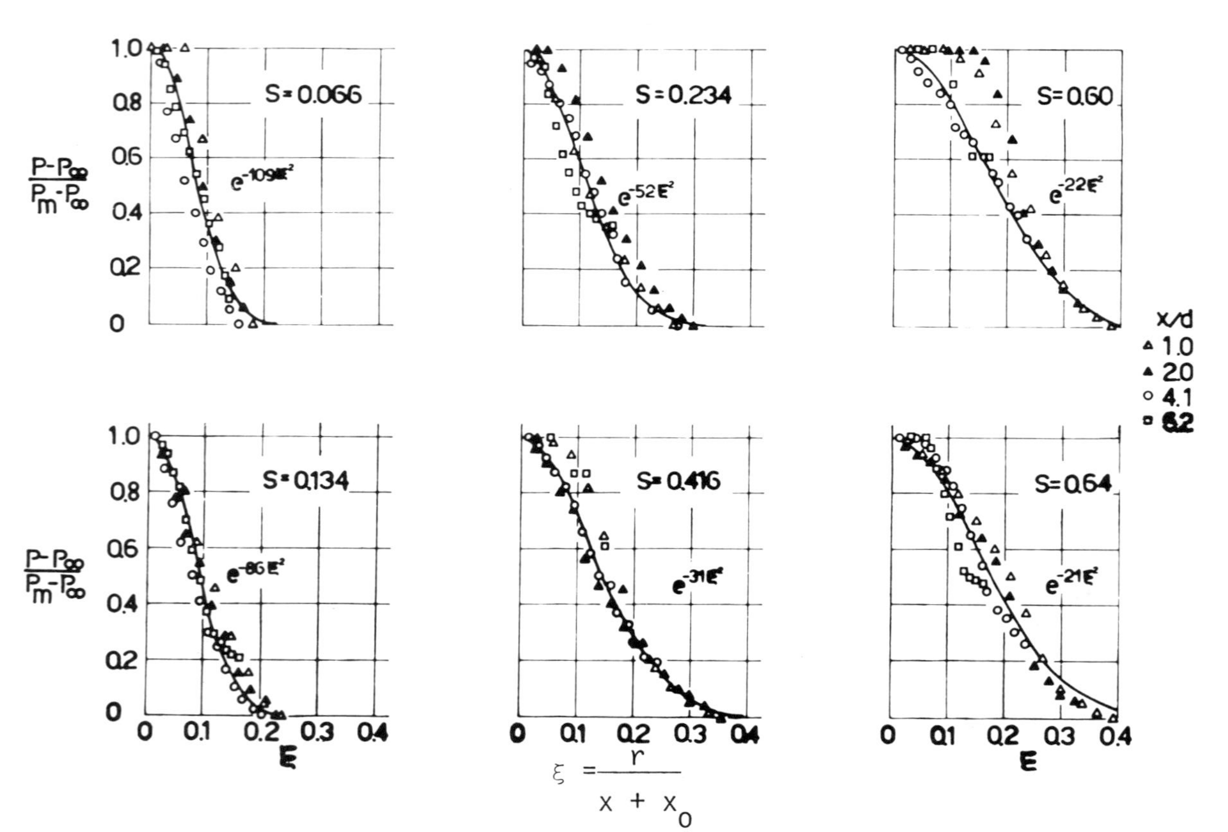

Fig. 110 **Radial profiles of static pressure in swirling jets as a function of swirl number[128] (published with permission).**

Eq. (13), is assumed to apply for cases with small swirl. No comparisons with experiment are provided.

In Ref. 18, an effort is made to generalize the mixing length model, Eq. (11), to a variety of new cases including flows with swirl. It is proposed that Eq. (14) be rewritten as

$$\nu_T = B(l'_m)^2 \, |\partial U/\partial y| \tag{126}$$

and l'_m be determined from

$$l'_m = (\Delta U)_{\max}/(\partial U/\partial y)_{\max} \tag{127}$$

with $B = 0.013$ for unswirling jets and $B = 0.017$ for swirling jets. A prediction of the largest value of Φ permissible before recirculation on the axis occurs, $\Phi = 0.60$, is in good agreement with experiment. No other comparisons with experiment are given.

An anisotropic eddy viscosity model is developed in Ref. 135. The rx shear is calculated from

$$\tau_{rx} = \rho\nu_{Tx}(\partial U/\partial r) \tag{128}$$

and the $r\theta$ shear from

$$\tau_{r\theta} = \rho\nu_{T\theta} r \frac{\partial}{\partial r}\left(\frac{W}{r}\right) \tag{129}$$

where

$$\rho\nu_{Tx} = \rho l_m^2 \left\{ \left(\frac{\partial U}{\partial r}\right)^2 + \left[r \frac{\partial}{\partial r}\left(\frac{W}{r}\right)\right]^2 \right\}^{1/2} \tag{130}$$

with

$$l_m = 0.08(1 + \lambda_s S_x) b \tag{131}$$

and

$$\rho\nu_{T\theta} = \rho\nu_{Tx}/\sigma_{r\theta} \tag{132}$$

The factors λ_s and $\sigma_{r\theta}$ are functions of the swirl rate through S_x, which is a "swirl" number with the same definition as the S we introduced before but now evaluated locally along the flow direction. This leaves two free parameters, λ_s and $\sigma_{r\theta}$, that can be used to obtain a "best fit" with experiment. The result of this process using the experiments of Ref. 128 is

$$\lambda_s = 0.6; \qquad \sigma_{r\theta} = 1 + 5.0 S_x^{1/3} \tag{133}$$

Comparison of the resulting predictions with the same data that were used to determine Eq. (133) is shown in Fig. 111.

2. *One-Equation Models*

For swirling jet flows, most workers have preferred to jump over the one-equation models and proceed directly to the two-equation models. Those efforts will be described in the next section.

A treatment based upon an integrated TKE equation model has been applied to a swirling wake case.[124,125] For cases with swirl, Eq. (80a) must be modified

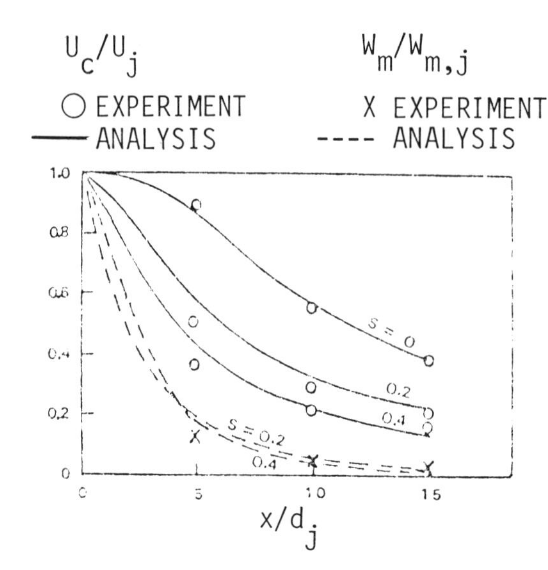

Fig. 111 Comparison of prediction with a mean-flow model and the experiment of Ref. 128 for the streamwise variation of axial and swirl velocity.[135]

to include the term in braces in Eq. (134):

$$\int_0^{b(x)} \frac{\partial}{\partial x}(\rho U k) r \mathrm{d}r = \int_0^{b(x)} -\rho \overline{uv}\left(\frac{\partial U}{\partial r}\right) r \mathrm{d}r + \left\{\int_0^{b(x)} -\rho \overline{wv}\left(\frac{\partial W}{\partial r} - \frac{W}{r}\right) r \mathrm{d}r\right\} - \frac{C_D}{l}\int_0^{b(x)} \rho k^{3/2} r \mathrm{d}r \tag{134}$$

Taking, in analogy with Eqs. (12) and (18),

$$-\rho \overline{wv} = \rho \nu_T \frac{\partial W}{\partial r} = \rho \sqrt{k} l \frac{\partial W}{\partial r} \tag{135}$$

an ordinary differential equation for $\nu_T(x)$ results:

$$\frac{\mathrm{d}\nu_T}{\mathrm{d}x} = \frac{C_3^2 U_\infty}{2}\left(\frac{\mathcal{I}_2}{\mathcal{I}_1}\right) + \left\{\frac{C_3^2 U_\infty}{2}\left(\frac{\mathcal{I}_3}{\mathcal{I}_1}\right)\right\} - \frac{C_D \nu_T^2}{4U_\infty C_3^2 b^2 \mathcal{I}_1} - \frac{\nu_T}{2\mathcal{I}_1}\frac{\mathrm{d}\mathcal{I}_1}{\mathrm{d}x} + \frac{\nu_T}{2b}\frac{\mathrm{d}\mathcal{I}}{\mathrm{d}x} - \frac{\nu_T}{2U_\infty}\frac{\mathrm{d}U_\infty}{\mathrm{d}x} \tag{136}$$

where

$$\mathcal{I}_3(x) \equiv \int_0^1 \left[\left(\frac{\partial(W/U_\infty)}{\partial(r/b)}\right)^2 - \frac{W/U_\infty}{r/b}\frac{\partial(W/U_\infty)}{\partial(r/b)}\right]\left(\frac{r}{b}\right)\mathrm{d}\left(\frac{r}{b}\right) \tag{137}$$

The first calculations made were for the plain, drag-body (propellerless) configuration, which, of course, had no swirl. Comparison with profiles from Ref. 121 at $x/D=2$ and 5 indicated a too rapid mixing rate [i.e., too high a $\nu_T(x)$]. Somewhat more direct evidence to support the contention that the predicted values of $\nu_T(x)$ were too high was found in the turbulence data. The three turbulence intensities were measured and reported, and this information can be used to estimate the average value of k. Using Eq. (18), one can then infer the predicted values of $\sqrt{k}$ to be compared with the measured values. This indicated that the predicted values were indeed too high, and reasonable variations in the constants C_3 and C_D did not solve the problem. Observing the downstream development of $\nu_T(x)$ obtained from Eq. (136), it was found that the strongest term that produced increases in $\nu_T(x)$ down over the body tail and into the wake was the first term on the right-hand side, which itself comes from the first term on the right-hand side of Eq. (134). As noted before, the shear term may be modeled with Eqs. (119), as suggested by Harsha,[82] rather than with Eq. (12), and this procedure gave good results for flow without a net momentum defect (see Sec. IV.C.3). For cases with swirl, it is now necessary to model a second component of the shear. Reference 125 adopted

$$-\rho\overline{wv} = A_\theta \rho k \tag{138}$$

with

$$A_\theta = a_\theta \frac{\partial W/\partial r}{|\partial W/\partial r|_{max}}; \qquad \frac{\partial W}{\partial r} > 0 \tag{139a}$$

or

$$A_\theta = a_\theta \frac{\partial W/\partial r}{|\partial W/\partial r_{min}|}; \qquad \frac{\partial W}{\partial r} < 0 \tag{139b}$$

with a_θ expected to be of order 0.3. All of this results, then, in a modified form of Eq. (136). Use of that equation produced better estimates of k_{av} and hence ν_T in comparison with experiment.

Next, calculations attempting a full simulation of the propeller-driven experimental case of Ref. 132 were run; i.e., swirl velocities were included. The simulation of the propeller is described in Ref. 125. The calculations were run with $a_\theta = 0.30$ by analogy with a_l. The axial velocity distribution obtained at $x/D = 2$ is compared with the data in Fig. 112. The prediction of swirl velocity is compared with the data in terms of flow angularity (pitch and yaw) at $x/D = 2$, as shown in Fig. 113. The calculations assume pure axisymmetric flow, so that the prediction for yaw angle for a horizontal traverse is identically zero, whereas the data show some small yaw angle. The prediction for pitch angle is qualitatively correct, but the absolute level of the maximum is too low. Slightly poorer agreement for the axial velocity was found in comparison with a prediction obtained neglecting swirl altogether. This is a result of the higher eddy viscosity values predicted for the calculation including swirl.

The effect of the value of a_θ was investigated by trying a calculation with $a_\theta = 0.15$. This resulted in lower eddy viscosity values, and a slightly improved axial velocity prediction, but somewhat lower pitch angles and poorer agreement with the swirl data. These results appear to indicate that a more elaborate model for turbulent transport including swirl may be required than that obtained by simple, direct extension of nonswirling models.

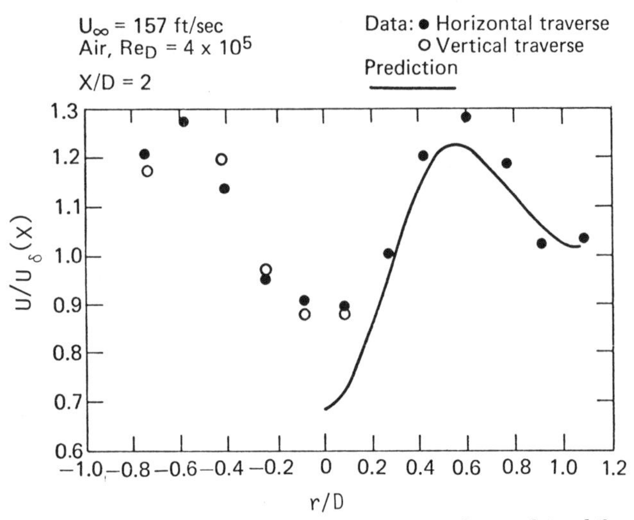

Fig. 112 Comparison of prediction with an integrated one-equation model and the experiment of Ref. 132 for the wake behind a propeller-driven body[125]: axial velocity profile.

3. *Two-Equation Models*

A model at this level is also developed in Ref. 135. A TKE equation, the same as the modeled form of Eq. (17) used in Ref. 136, but for the added swirl term on the right-hand side shown in Eq. (140), was employed:

$$\rho\left(U\frac{\partial k}{\partial x}+V\frac{\partial k}{\partial r}\right)=\frac{1}{r}\frac{\partial}{\partial r}\left(r\frac{\rho\nu_{Tx}}{\sigma_k}\frac{\partial k}{\partial r}\right)+\rho\nu_{Tx}\left(\frac{\partial U}{\partial r}\right)^2 -C_D\frac{\rho k^{3/2}}{l}+\rho\nu_{T\theta}\left[r\frac{\partial}{\partial r}\left(\frac{W}{r}\right)\right]^2 \tag{140}$$

The Z equation employed was that derived from the choice $\alpha=\beta=1$, i.e., $Z\equiv kl$, following directly from the work of Ref. 136, again with a term added for swirl:

$$\rho\left(U\frac{\partial(kl)}{\partial x}+V\frac{\partial(kl)}{\partial r}\right)=\frac{1}{r}\frac{\partial}{\partial r}\left(r\frac{\rho\nu_{Tx}}{\sigma_{kl}}\frac{\partial(kl)}{\partial r}\right) +C_Bl\left\{\rho\nu_{Tx}\left(\frac{\partial U}{\partial r}\right)^2+\rho\nu_{T\theta}\left[r\frac{\partial}{\partial r}\left(\frac{W}{r}\right)\right]^2\right\}-C_S\rho k^{3/2}+C_R\rho k^{3/2}(Ri) \tag{141}$$

where Ri is a Richardson number defined as

$$Ri\equiv\frac{2(W/r^2)\partial(rW)/\partial r}{(\partial U/\partial r)^2+[r\partial(W/r)/\partial r]^2} \tag{142}$$

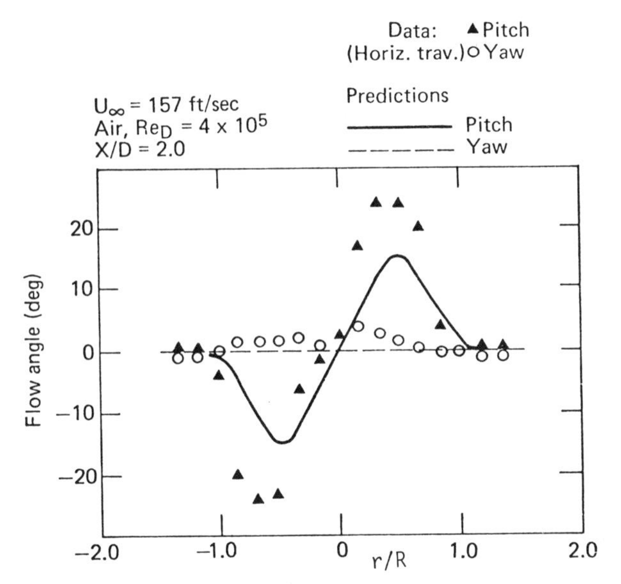

Fig. 113 Comparison of prediction with an integrated one-equation model and the experiment of Ref. 132 for the wake behind a propeller-driven body[125]: flow angularity.

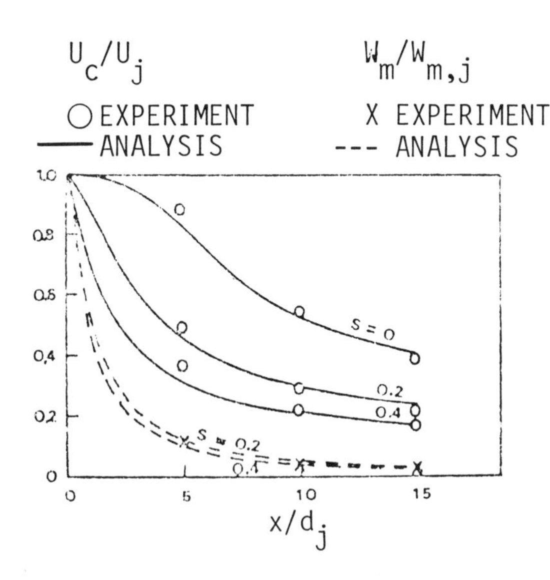

Fig. 114 Comparison of prediction with a two-equation model and the experiment of Ref. 128 for the streamwise variation of axial and swirl velocity.[135]

The constants $C_D = 0.055$, $C_B = 0.98$, $C_S = 0.0397$, $\sigma_k = 1.0$, and $\sigma_{kl} = 1.0$ were taken directly from Ref. 136, and C_R was determined by comparison with a swirling jet experiment.[128] The eddy viscosities were determined from

$$\rho \nu_{Tx} = \rho \sqrt{k} l \tag{143}$$

and Eq. (132). Computer optimization using the experiments led to

$$C_R = 0.06; \qquad \sigma_{r\theta} = 1 + 2.0 S_x^{1/3} \tag{144}$$

and comparison of the predictions with the same data used to determine Eqs. (144) is shown in Fig. 114.

Rodi[137] has worked with an extended $k\epsilon$ model for swirl problems. His formulation employed an isotropic eddy viscosity, i.e., Lilley's $\sigma_{r\theta} \equiv 1.0$. An attempt was made to vary the proportionality constants to obtain good agreement with one experiment and then make predictions for other experiments. This was not successful. In general, too slow a decay of the swirl velocity was obtained.

4. *Reynolds Stress Models*

Reference 104 presents an in-depth study of swirl problems with this type of model. The discussion of the work of Ref. 104 for non-swirling problems at the end of Sec. II.C.8 is important for that which follows here. Using the basic formulation of Ref. 103 written in axisymmetric coordinates and with the terms due to a swirling motion added, predictions were made and compared to the data of Ref. 129. The results were very discouraging. The inclusion of the equation for $\overline{uw}$ actually resulted in a predicted decrease in mixing rate due to swirl as opposed to the increase always observed in experiment. The calculated values of $\overline{uw}$ were predominantly negative, which resulted in a predicted decrease in $\overline{uv}$. From experiment, $\overline{uw}$ is generally found to be positive. It was found that modification of the pressure-strain correlation on an ad hoc basis did not improve the situation markedly. This led to the conclusion that the use of the more elaborate model of Ref. 138 for these processes is probably necessary. That has not, as yet, been accomplished. Finally, this work casts doubt upon the soundness of the turbulence modeling in Ref. 135.

VI. TWO-PHASE FLOWS

A. Introduction

There are many situations of practical interest in which the mixing of fluid streams where one or both contain two phases occurs. Two examples are atomizing liquid jets in an airstream and a soot-laden exhaust from a smoke stack. These two involve liquid droplets in a gas and solid particles in a gas. Other examples would have solids in a liquid stream and gas bubbles in a liquid stream. Therefore, the actual scope of the problems of practical importance is very large. The situation is complicated further by the fact that "particles" over a very wide size (the chart in Fig. 1.1 of Ref. 139 is illuminating) and weight range are of interest and also the fact that many problems involve a distribution of particle sizes. Finally, many engineering cases are concerned with "particles" that are not simple spheres or indeed any simple, regular shape, and they may also deform in the flow. With all of this, it should not surprise the reader to learn that the present state of knowledge in this general area is rather spotty and, on the average, poor. Indeed, there remains considerable disagreement as to some of the gross effects involved.

The primary emphasis here will be on particles (either solid or liquid) in a fluid stream (either liquid or gas). No such discussion would be complete without some mention of the common methods by which the particle (or particle-laden) stream is generated. Small solid particles are generally produced external to the flow system by some mechanical crushing or grinding operation, whereas liquid particles are often produced from the breakup of a liquid stream by an atomization device that is a part of the flow system proper. The closest case to the latter for solids in a fluid stream is particle pickup of coal dust by an airstream. There, fine dust is observed to go into suspension in the airstream by first breaking loose in large chunks which subsequently break up. Since solid particles are usually produced external to the flow system, more will not be said about those processes here. However, the methods of introduction of, for example, a powder into a gas stream are worthy of mention. The two most common methods are direct injection by means of a rotating screw in a tube or a fluidized bed vented into the stream.

Because atomization of liquid injectants is commonly a direct part of the total flow system, a short introduction to the most common types of atomizers is included here. The job of the atomizer is to increase the surface area of a given volume rate of liquid, so as, for example, to enhance the rate of surface heat and mass transfer. For perspective, the atomization of 1 cm^3 of liquid into drops of 100 μm diam (a value in the range of practical interest) increases the surface area by a ratio of more than 1200:1. Industrial atomizers for use with liquid fossil fuels must be designed for cases covering a range in viscosity of nearly a 100:1.

The simplest type of atomizer, known as the "pressure jet" type, is nothing more than a high-velocity liquid jet produced by an orifice with a high back pressure. Such a high-velocity jet is basically unstable to small random disturbances, and it breaks up of its own accord. "Swirl" atomizers produce a hollow cone or flat sheet by pressurized injection through an orifice with an initial tangential velocity component in the liquid stream. These sheets are also very unstable. These two types of device are used to produce drops in the 100-200-μm range. "Assisted pressure jet" atomizers employ a secondary stream of high-velocity gas (usually air or steam) injected into the early stages of the spray cone. This reduces the drop size somewhat and provides better operation over a range of injectant flow rate. "Blast" atomizers rely on the interaction of a high-speed gas stream (often the main stream itself) and a relatively slow liquid jet. The simplest type, the "air blast" atomizer, has a transverse liquid jet in a gas stream, and a more extensive discussion of that important flow situation will be given here in Chapter VIII on Transverse Injection. The blast atomizers are used to produce sprays in the 40-80-μm range. In the "rotary atomizer," the fluid is fed onto a rotating disk or hollow cup, and the atomization is achieved by centrifugal force. These have the advantage of being insensitive to the viscosity of the liquid. The last common type is the "impinging jet" atomizer where two (or more) high-pressure jets are arranged to intersect to accelerate breakup. In liquid-fuel rockets, one can have fuel jets intersect, or one can have pairs of fuel and liquid oxidant jets intersect to enhance mixing as well as breakup. This type of atomizer produces the smallest drops.

In this volume, the discussion will be limited to cases with non-deformable and uncharged particles of spherical shape, since the best and most general work available is for such cases. The sections on experimental data and analysis are divided into subsections on single particles in a turbulent flow and jet mixing cases.

B. Experiments

1. Single Particle in a Turbulent Flow

Most of the work in this field begins with the drag coefficient for a sphere. Of course, this information is well established for cases with large spheres in low-turbulence streams; however, the main interest here is for small spheres in turbulent streams where the scale of the turbulence may be large compared to the particle (sphere) size. Other complicating factors are also important in the context of two-phase flow problems. The particles are, in general, accelerating or decelerating and rotating; they often are moving in a local pressure gradient, and there may be a temperature difference between the particle and the fluid stream. If there is evaporation, sublimation, or condensation on the particle surface, the problem becomes vastly more complex. In spite of all of this, most workers rely upon a standard C_D vs Re_d experimental curve, e.g., Fig. 2.1 in Ref. 139, for lack of anything better.

The parameter that determines the gross behavior of a particle in turbulent flow is generally taken as the ratio of the particle relaxation time τ_s to the Lagrangian macrotimescale T_l. For $(\tau_s/T_l) \ll 1$, the particle will completely follow the turbulent flow, and, for $(\tau_s/T_l) \approx 1$, the particle cannot follow the higher-frequency turbulent motions. Unfortunately, direct measurements of this important parameter over a range of the important flow variables are scarce. A discussion is given in Chapter 2 of Ref. 139.

2. *Clouds of Particles in Turbulent Flow*

The simplest question to be dealt with here concerns how large the concentration of particles must be before the particles cease to behave as isolated particles. A limit of 2% concentration of particles by volume is suggested in Ref. 139. The variation of C_D vs Re_d as a function of ϵ_p ($\equiv 1$ volume fraction of particles) is shown here in Fig. 115 taken from Ref. 139.

The motion of a cloud of particles may also be expected to influence the turbulence of the fluid stream, as well as vice versa. Measurements of these processes in duct flows are presented and discussed in some detail in Chapter 2 of Ref. 139.

3. *Particle-Laden Jets*

A series of experiments with particle-laden jets exhausting into stationary surroundings have been conducted by Goldschmidt and co-workers and summarized in Ref. 140. In Fig. 116, we show the axial variation of centerline concentration flux of the contaminant DBP (dibutyl phthalate) for the case of DBP droplets in a planar air jet for various droplet sizes. Transverse profiles are shown in Fig. 117, and the concentration flux half-width is given in Fig. 118. Taken together, these data imply the surprising result that mixing is accelerated by the larger particles. The fact that the mixing zone is found to be wider for the larger particles is especially troubling. Similar results were, however, found for N_2 bubbles in a circular water jet.

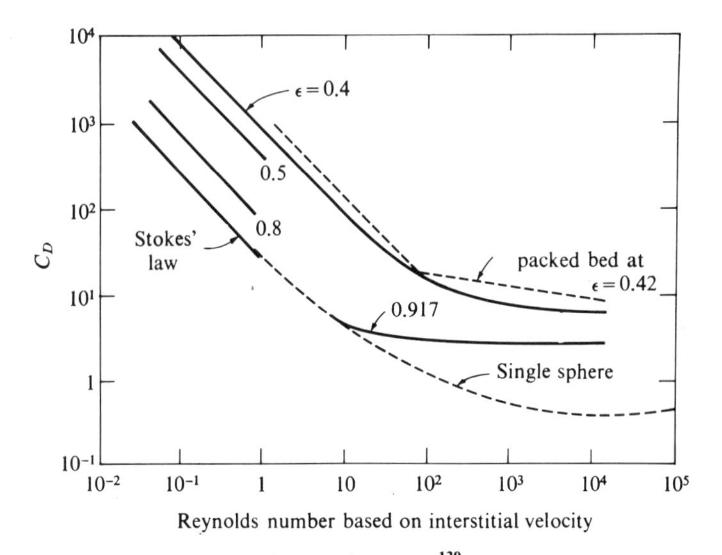

Fig. 115 Drag coefficient of spheres in a cloud[139] (published with permission).

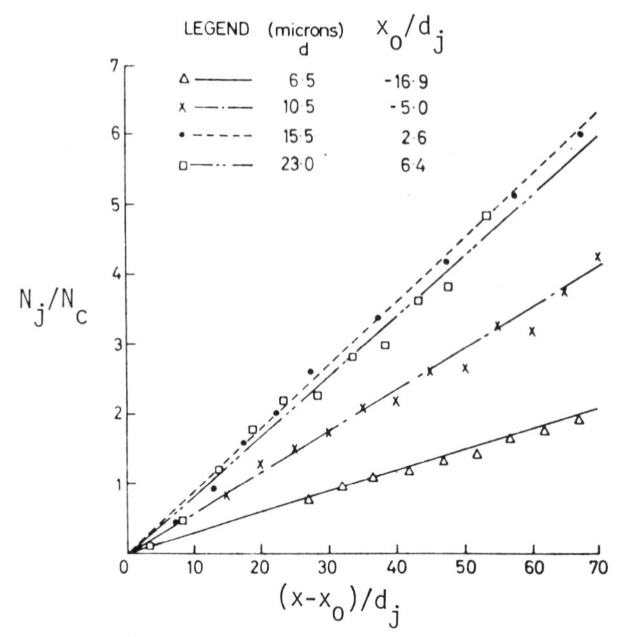

Fig. 116 Streamwise variation of concentration flux of particles (N = counts/time) for DBP droplets in a planar air jet[140] (published with permission).

For the N_2-water experiments, it was found that the mean velocity profiles, axial velocity decay, and velocity half-width growth were all essentially unaffected by the presence of the bubbles compared to the same tests without bubbles. The turbulent intensity in the water stream is decreased slightly by the presence of the bubbles, as shown in Fig. 119. The energy spectrum also showed a small effect of the bubbles, as may be seen in Fig. 120.

An interesting series of experiments is presented in Ref. 141. Here, we have a planar air jet into nominally quiescent surroundings. The air that is entrained into one side of the jet was seeded separately with gaseous CO and MgO particles in two sizes (1 and 9 μ) in different tests. The main results are given in terms of the influence of particle size on turbulent Schmidt number S_{cT} [see Eq. (52)]. The value of $S_{cT} = 0.34$ found in this work is considerably lower than that generally accepted for gas-gas mixing (see Sec. II.B.5). The tests for the small MgO particles yielded $S_{cT} = 0.34$ (the same as for the CO-air test), but the larger MgO particles resulted in $S_{cT} = 0.17$. The results of Ref. 140 similarly indicated a decrease in S_{cT} with increasing particle size. This can be interpreted to mean that the eddy diffusivity increases with particle size, a result that many workers find hard to accept on intuitive grounds.

Experiments on the mixing of a helium/air jet (both with and without particles) in a secondary airstream were reported in Ref. 142. The particles were 6- and 30-μ aluminum spheres. The influence of particles and particle size is shown in Fig. 121 in terms of the decay of helium centerline concentration.

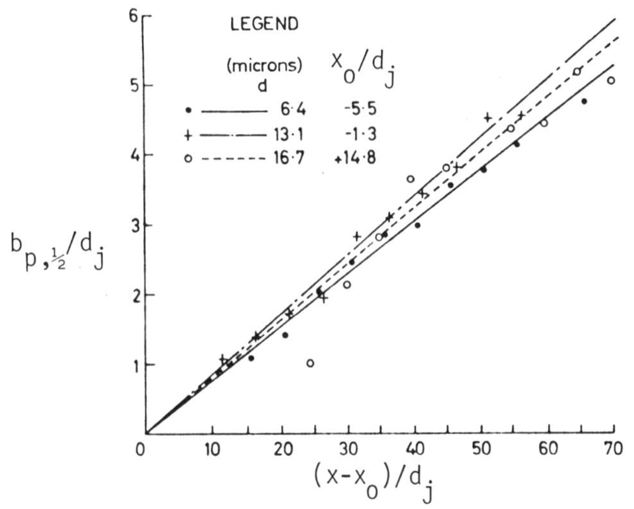

Fig. 117 Variation of concentration flux half-width for DBP droplets in a planar air jet[140] (published with permission).

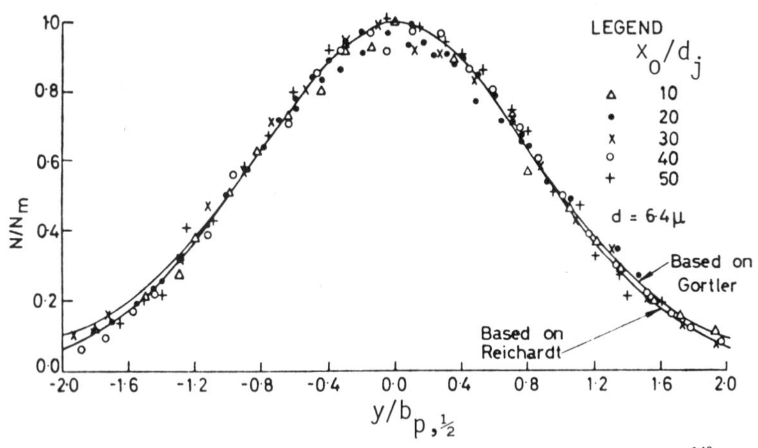

Fig. 118 Transverse concentration flux profiles for DBP droplets in a planar air jet[140] (published with permission).

The systematic experiments of Refs. 143-145 provide perhaps the best information available. The tests were conducted with a round air jet laden with corundum powder of various sizes (17, 32, 49, 72, and 80 μ) with various initial loadings exhausting into still air. Particular attention was paid to the initial uniformity of the particles in the jet. The effect of particles on the axial decay of the centerline velocity is shown in Fig. 122, where it can be seen that, the smaller the particles, the greater will be the decrease in decay rate. Each experiment is characterized by a pair of numbers indicating particle size and

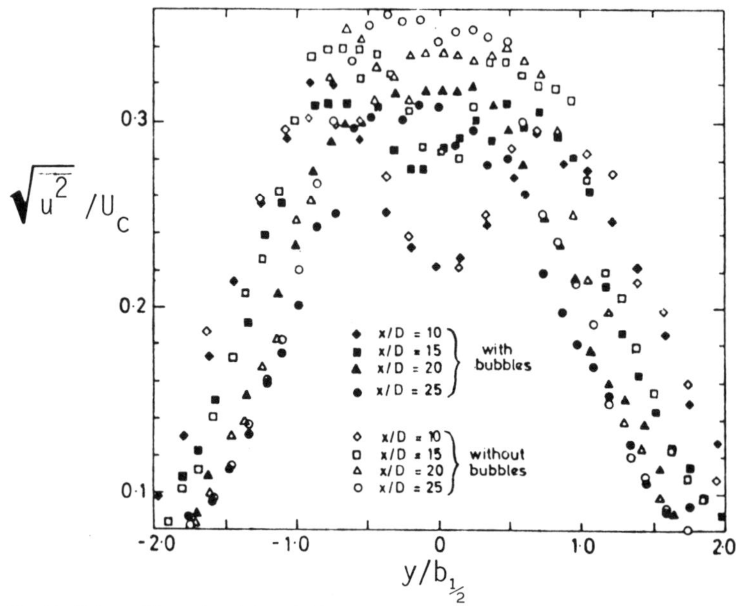

Fig. 119 Radial profiles of axial turbulence intensity in an axisymmetric water jet with N_2 bubbles[140] (published with permission).

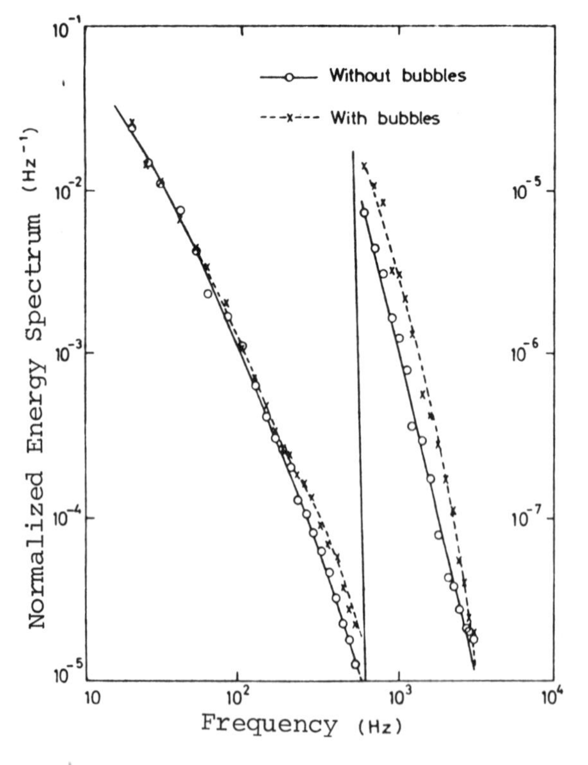

Fig. 120 Energy spectra in an axisymmetric water jet with N_2 bubbles[140] (published with permission).

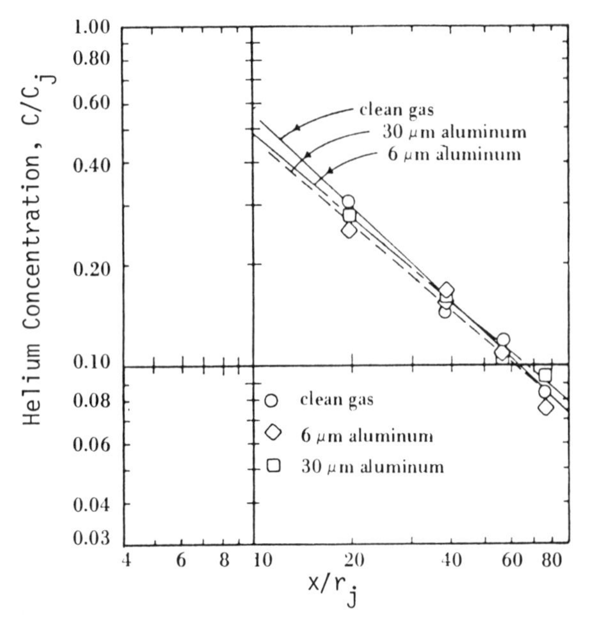

Fig. 121 Axial variation of centerline injectant (helium) concentration for an axisymmetric jet in a coaxial airstream for various particle sizes in the jet flow [142] (published with permission).

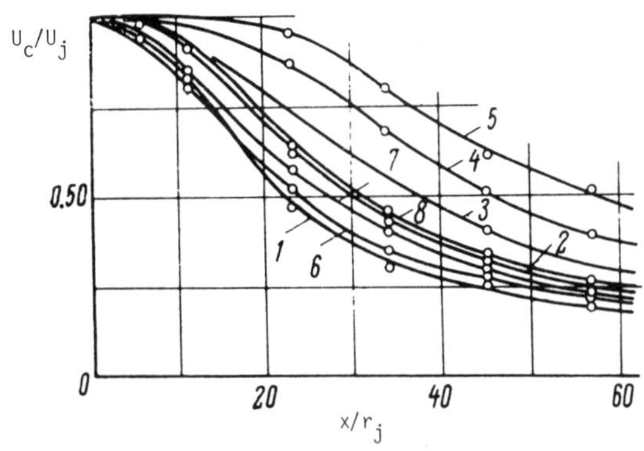

Fig. 122 Streamwise variation of centerline velocity in a particle-laden axisymmetric jet [145]: 1(all gas), 2(32, 0.3), 3(32, 0.56), 4(32, 0.77), 5(32, 1.4), 6(72, 0.3), 7(49, 0.3), 8(17, 0.56) (published with permission).

initial loading, and the curves numbered 2 through 8 correspond to, for example, 2(32, 0.3), with curve 1 being the pure air jet case. The conclusion stated earlier can be seen by comparing curves 2, 6, and 7 or 3 and 8. For a given particle size, the greater the loading, the greater will be the decrease in decay rate; compare curves 2, 3, 4, and 5.

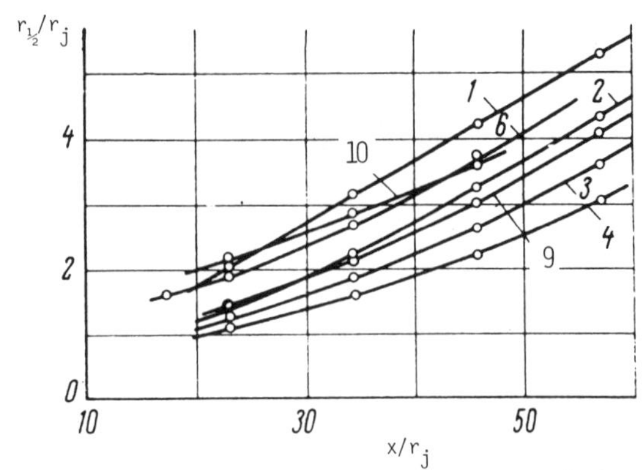

Fig. 123 Streamwise growth of velocity half-radius in particle-laden axisymmetric jets[145]: 1(all gas), 2(32, 0.3), 3(32, 0.56), 4(32, 0.77), 6(72, 0.3), 9(17, 0.3), 10(80, 0.3) (published with permission).

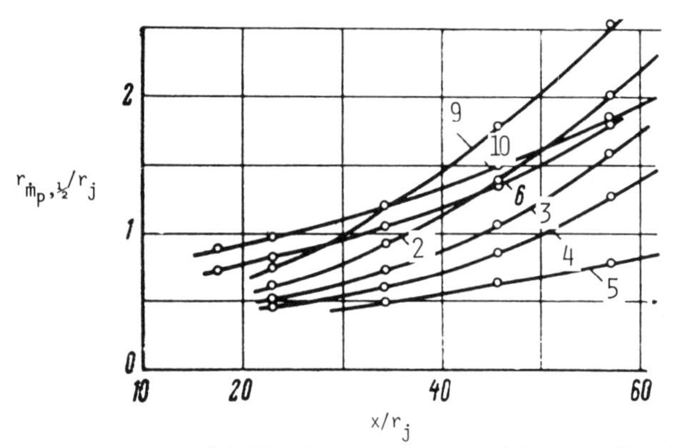

Fig. 124 Streamwise growth of half-radius based on particle mass flux in particle-laden axisymmetric jets[145]: 2(32, 0.3), 3(32, 0.56), 4(32, 0.77), 5(32, 1.4), 6(72, 0.3), 9(17, 0.3), 10(80, 0.3) (published with permission).

The axial growths of the half-radius based on velocity and the half-radius based on flow-rate are shown in Figs. 123 and 124, respectively. Taken together, these results show that, the finer the additive, the more rapidly the slope of the half-radius based on velocity approaches that for the pure air jet, and the more rapidly the effect of the additive decays. Furthermore, near the exit, the coarser particles appear to diffuse more rapidly, but further downstream the consistent pattern of more rapid diffusion of the finer additives emerges. Finally, an increase in additive concentration leads to a decrease in jet expansion and diffusion.

Many of the available experimental data have been correlated in Ref. 146 for the mean fluid velocity and the particle mass flux fields in terms of the initial loading of the particles. Two main regions have been found to exist: a near field where no momentum has been transferred between the phases, and a far field in which sensibly all the momentum is in the fluid phase.

C. Analysis

1. *Single Particle in a Turbulent Flow*

A considerable number of studies of this special limiting case have been undertaken, and Refs. 139 and 147 contain descriptions of most of them. Since our main interest here is with clouds of particles in turbulent shear flows, only a brief introduction to this subject will be presented. The discussion has been taken largely from Ref. 147.

In a turbulent field, the particle is subjected to time-varying forces by the surrounding flow. The important parameter here is the response half-time that was mentioned earlier. It is possible with a little analysis to describe that quantity and a characteristic time for the fluid eddies more precisely. We shall work with a simplified form of the general equations for the dynamics of a single particle,[147] and consider the drag due to relative particle and fluid motion as the principal force on the particle. Moreover, the treatment is based on a quasisteady assumption for the drag law. The steady resistance may be written as

$$f(Re_s)\,d_s \mu U_r \tag{145a}$$

with

$$Re_s \equiv \frac{\rho_s d_s\,|U_r|}{\mu} \tag{145b}$$

The function $f(Re_s)$ is simply a constant in the Stokes' flow regime, d_s is the dimension of the particle, and $U_r \equiv (U_s - U)$ is the relative velocity. With all of this, the equation for the motion of the particle can be written as

$$\frac{\mathrm{d}(Re_s)}{\mathrm{d}t} + \frac{\mu}{(\rho_s + \Omega\rho)\,d_s^2} f_l\,(Re_s) = \cdots \tag{146}$$

where $f_l\,(Re_s) = Re_s \cdot f(Re_s)$, and $\Omega\rho\cdot$(volume of a particle) is the virtual mass. Let

$$F(Re_s) \equiv \int \frac{\mathrm{d}(Re_s)}{f_l\,(Re_s)} \tag{147}$$

and the homogeneous part of the solution for $F(Re)$ is

$$F[Re_s(t)] = F[Re_s(t=0)] - \frac{\mu}{(\rho_s + \Omega\rho)\,d_s^2}\cdot t \tag{148}$$

The particle half-time τ_s is then

$$\tau_s = \frac{F[Re_s(t=0)] - F[\tfrac{1}{2}Re_s(t=0)]}{\mu}\, d_s^2 \cdot (\rho_s + \Omega\rho) \tag{149}$$

For Stokes' flow, $f_l\,(Re_s = K_l \cdot Re_s)$, with $K_l \approx O(10)$ and

$$F(Re_s) = (1/K_l)\,\ln(Re_s) \tag{150}$$

so that

$$\tau_s = \frac{\ln(2)}{K_1} \frac{(\rho_s + \Omega\rho)}{\mu} d_s^2 \tag{151}$$

The characteristic time of an eddy may be taken as $(uK_s)^{-1}$, where $K_s = 1/d_s$ is the wave number for an eddy of size d_s. The rms turbulence intensity u can be taken as

$$u(K_s) = [K_s \cdot E(K_s)]^{1/2} \tag{152}$$

Thus, the characteristic time of the fluid turbulence is

$$[K_s^3 \cdot E(K_s)]^{1/2} \tag{153}$$

This can be recast in terms of the dissipation ϵ as

$$(\nu/\epsilon)^{1/2} \tag{154}$$

and the ratio of particle to eddy characteristic times becomes

$$\frac{\tau_s}{(\nu/\epsilon)^{1/2}} \propto \left(\frac{\rho_s}{\rho} + \Omega\right) \cdot O\left(0.1 \frac{d_s^2}{\lambda^2}\right) \tag{155}$$

where $\lambda = (\nu^3/\epsilon^4)^{1/4}$ is Kolmogoroff microscale. Therefore, for $\rho_s \approx \rho$, if we have $d_s \approx \lambda$, the particles can follow the smallest eddies. If $\rho_s \gg \rho$, then $d_s \ll \lambda$ must be satisfied for that condition to be achieved. Of course, most practical cases fall somewhere in between.

2. *Mean-Flow Models*

The early analyses[5,139-141] were based on simple extensions of gaseous, variable composition treatments. In these, any direct effects of the particles on the gas-phase turbulence was essentially ignored. Stated most simply, the effect of the particles was contained only in the value of the turbulent Schmidt number in the presence of particles. The American work (e.g. Refs. 140 and 141) generally showed a decrease in Schmidt number with an increase in particle size. If the gas-phase flow is unaffected by the particles, this result indicates an increase in diffusion rate with an increase in particle size. This is contrary to physical intuition and to the results of Ref. 145. It should perhaps be emphasized here that the data reported do show very considerable scatter and uncertainty for S_{cT}.

The results of Ref. 145 have prompted a series of new analyses in the Soviet Union.[148-150] The treatment is based upon the boundary-layer form of the "two-fluid" equations of motion:

$$\frac{\partial}{\partial x}(Uy^j) + \frac{\partial}{\partial y}(Vy^j) = 0 \tag{156}$$

$$U\frac{\partial U}{\partial x} + V\frac{\partial U}{\partial y} = -\frac{1}{y^j}\frac{\partial}{\partial y}(\overline{uv}y^j) + f_x \tag{157}$$

$$\rho_s U_s \frac{\partial U_s}{\partial x} + (\rho_s V_s + \overline{\rho_s' v_s})\frac{\partial U_s}{\partial y} = -\frac{1}{y^j}\frac{\partial}{\partial y}(\rho_s \overline{u_s v_s} y^j) \tag{158}$$

$$\frac{\partial}{\partial x}(\rho_s U_s y^j) + \frac{\partial}{\partial y}[(\rho_s V_s + \overline{\rho_s' v_s}) y^j] = 0 \tag{159}$$

Here, the subscript s denotes the particle phase, and f_x is the longitudinal component of the mean interaction force between the two phases. Note that these equations and all those that follow have been made dimensionless using U_j for velocities, r_j for lengths, and the gas-phase density for densities. Therefore, ρ_s is actually the particle concentration. We can see that the motion of the particles is determined by two factors: the interaction force f_x, and the apparent tangential stress $\rho_s(\overline{u_s v_s})$.

The interaction force is modeled using the drag law for a sphere with

$$f_x = \beta G \rho_s (U_s - U) \tag{160}$$

where

$$\beta \equiv \frac{r_j}{U_j \tau}; \qquad \tau \equiv \frac{\rho_s^0 d_s^2}{18\mu}; \qquad G = C_D \frac{Re}{24} \tag{161a}$$

$$Re = \frac{\rho |U - U_s| d_s}{\mu} = \frac{\rho}{\rho_s}(Re_s) \tag{161b}$$

and where ρ_s^0 is the density of the particle material and $G(Re)$ is the departure from Stokes drag law, for which $G \equiv 1$.

The main new treatment concerns the modeling of the correlation terms. This is begun with some standard concepts:

$$-\overline{uv} = \nu_T \frac{\partial U}{\partial y}; \qquad -\overline{u_s v_s} = \nu_{TS} \frac{\partial U}{\partial y} \tag{162a}$$

$$-\overline{\rho_s' \nu_s} = \frac{\nu_{TS}}{Sc_T} \frac{\partial \rho_s}{\partial y} \tag{162b}$$

$$\nu_T = \text{const } |v| l_m; \qquad \nu_{TS} = \text{const } |v_s| l_m \tag{163}$$

The results of Ref. 144 have been interpreted in Ref. 148 to yield

$$|v| = |v_0| \frac{1 + \rho_s (v_r / v_0)}{1 + \rho_s} \tag{164a}$$

$$|v_s| = |v_0| \frac{1 - (v_r / v_0)}{1 + \rho_s} \tag{164b}$$

where $v_r \equiv v - v_s$. In deriving Eqs. (164), it was assumed that a turbulent lump with initial transverse velocity $v_0 = l_m (\partial U / \partial y)$ is impeded by the particles whose initial velocity is taken as zero. As a result, after moving a distance l_m, the velocity decreases as given by the equations. Using these expressions in Eqs. (163) leads to

$$v_T = \text{const } l_m^2 \left[\frac{1 + \rho_s (v_s / v_0)}{1 + \rho_s} \right] \left| \frac{\partial U}{\partial y} \right| \tag{165a}$$

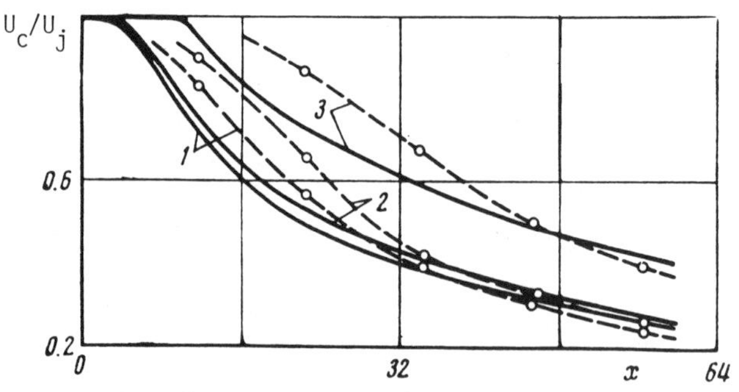

Fig. 125 Comparison of prediction and experiment for centerline velocity of particle-laden axisymmetric jets[150]:------experiment from Ref. 145; ———prediction; 1(49, 0.3), 2(32, 0.3), 3(32, 0.77) (published with permission).

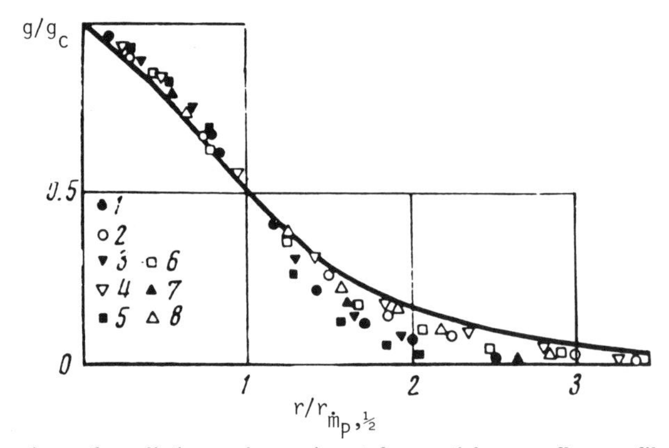

Fig. 126 Comparison of prediction and experiment for particle mass flux profiles in particle-laden axisymmetric jets[150]: ———experiment from Ref. 145; □ ■ ● ○ ▲ △ ▼ ▽ predictions for 1(32, 0.3; $x = 22.2$), 2(32, 0.3; $x = 55.5$), 3(32, 0.77; $x = 22.2$), 4(32, 0.77; $x = 55.5$), 5(49, 0.3; $x = 55.5$), 6(49, 0.3; $x = 55.5$), 7(17, 0.3; $x = 22.2$), 8(17, 0.3; $x = 55.5$) (published with permission).

$$v_{TS} = \text{const } l_m^2 \left[\frac{1-(v_s/v_0)}{1+\rho_s} \right] \left| \frac{\partial U}{\partial y} \right| \tag{165b}$$

The constants in Eqs. (165) were taken as equal and as that for all-gas mixing. With the mixing-length determined from

$$l_m = \Delta U / \left| \partial U/\partial y \right|_{\max} \tag{166}$$

the constant was chosen as 0.013.

The equations resulting from this modeling were solved numerically in Ref. 150 for the conditions of the experiments in Ref. 145. It was necessary to assume a value of the initial particle velocity in comparison to U_j; it was taken as 80% of U_j. Furthermore, Sc_T was taken as constant at 1.6, which is in rough agreement with the variation as a function of β and ρ_s predicted in Ref. 149.

The predictions (solid lines) are compared with the data (dashed lines) for three cases in Fig. 125. The qualitative ordering of mixing rate as a function of particle size and loading is predicted correctly. Also, the quantitative agreement between prediction and experiment is quite good for the larger values of x where the influence of the details of the injection process (i.e., particle distribution, etc.) can be expected to be small. The calculations also correctly predicted the "similarity" of the specific particle flow rate for $x > 20$, as shown in Fig. 126.

3. *Higher-Order Models*

None of the higher-order models have been applied to two-phase flow problems.

VII. THREE-DIMENSIONAL, COAXIAL JETS

A. Scope

Three-dimensional mixing problems generally occur in practice in one of three ways. First, the injector nozzle may itself be three-dimensional, e.g., rectangular or elliptical. Second, there may be more than one injector exhausting parallel to a mainstream, but the injectors are close enough that the mixing zones from each interfere. Third, the injector may exhaust at an angle to a mainstream. The first two cases are coaxial and will be discussed in this chapter. The third situation forms the basis of a later, separate chapter.

B. Results from Experiment

1. Three-Dimensional Nozzles

A series of experiments has been published for three-dimensional nozzles exhausting into quiescent surroundings.[151-154] The streamwise decay of the velocity along the axis for several cases from Refs. 151 and 153 is shown in Fig. 127, where $e^* \equiv d/l_l$, the ratio of the minor to the major axis of the orifice, where relevant. Each orifice had the same area, and the initial momentum was within a 12% spread for all jets. In order to describe cases of this type in a reasonably uniform manner, the three-region flow picture shown in Fig. 128 was constructed in Refs. 151 and 153.

1) Potential core (PC) region: The mixing initiated at the boundaries has not yet permeated the entire flowfield, leaving a region of uniform velocity close to the jet exit velocity.

2) Characteristic decay (CD) region: The axis velocity decay is dependent upon orifice configuration, and velocity profiles in the plane of the minor axis of the orifice are found to be "similar," whereas those in the plane of the major axis are "nonsimilar."

3) Axisymmetric decay (AD) region: The axis velocity decay is axisymmetric in nature, i.e., $U_c \sim x^{-l}$, and the entire flow is found to approach axisymmetry. Far downstream, a fourth region of fully axisymmetric flow is observed. The extent of these various regions is shown in Fig. 129.

The growth of the half-widths in both transverse directions is given in Fig. 130 for two rectangular orifice cases, along with that for an axisymmetric jet. For slender jets ($e^* \ll 1.0$), it was found that all of the results could be normalized to one set of curves if both the ordinate and the abscissa of Fig. 130 were multiplied by $(e^*/e^*_{\rm ref})$, where $e^*_{\rm ref}$ was taken as 0.10.

The transverse velocity distributions obtained for slender jets display "irregularities," as can be seen in Fig. 131. These were found not to be due to any nonuniformities in the jet exit profile or to the details of the nozzle shape ahead of the orifice. The irregularities were traced to the influence of a three-dimensional vortex "ring" that issues from the jet exit.

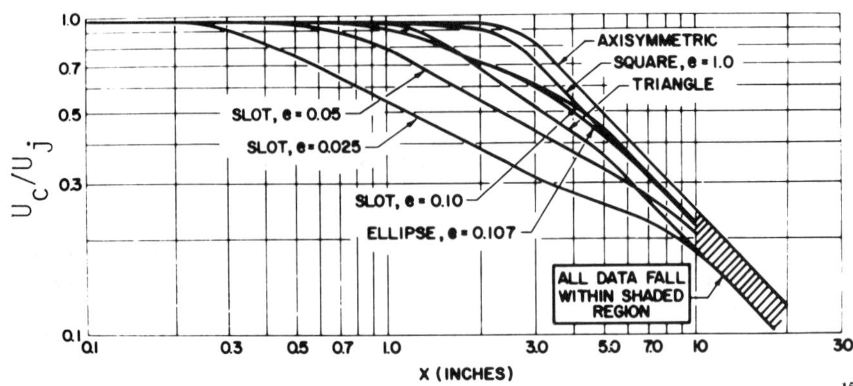

Fig. 127 Streamwise variation of axis velocity for several three-dimensional jets with $U_\infty = 0$.[151]

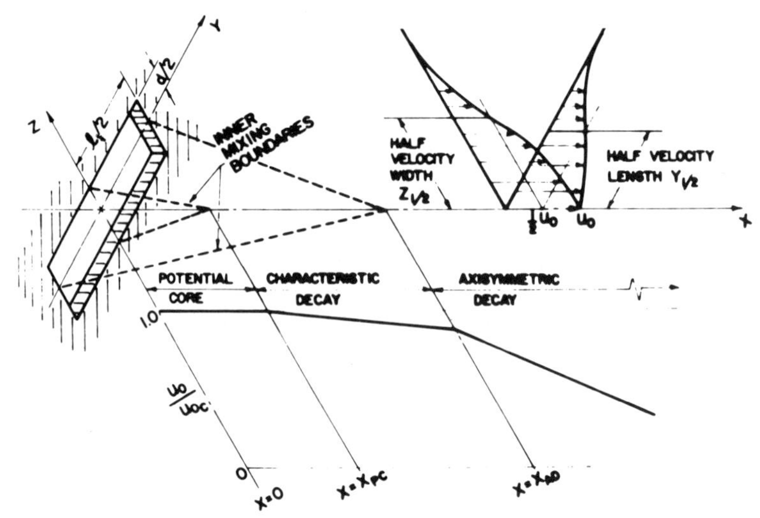

Fig. 128 Schematic representation of the flowfield for three-dimensional jets showing the three regions.[151]

2. *Adjacent, Coaxial Jets*

The studies reported in Ref. 155 were undertaken to investigate the effects of injection from a cluster of four coaxial jets into a supersonic $(M_\infty = 4)$ airstream. The injectant was helium, and the main parameter varied was the interjet spacing s/d_j. Figure 132 is a plot of the maximum helium concentrations (disregarding their actual radial locations, although generally this was along the original axis of the jet) as a function of axial distance for different (s/d_j). As seen in this graph, s/d_j has a somewhat inconsistent effect on the concentration at lower (x/d_j), but starting at $x/d_j = 36$ the influence of s/d_j is clear and as might be expected: lower concentration at higher s/d_j.

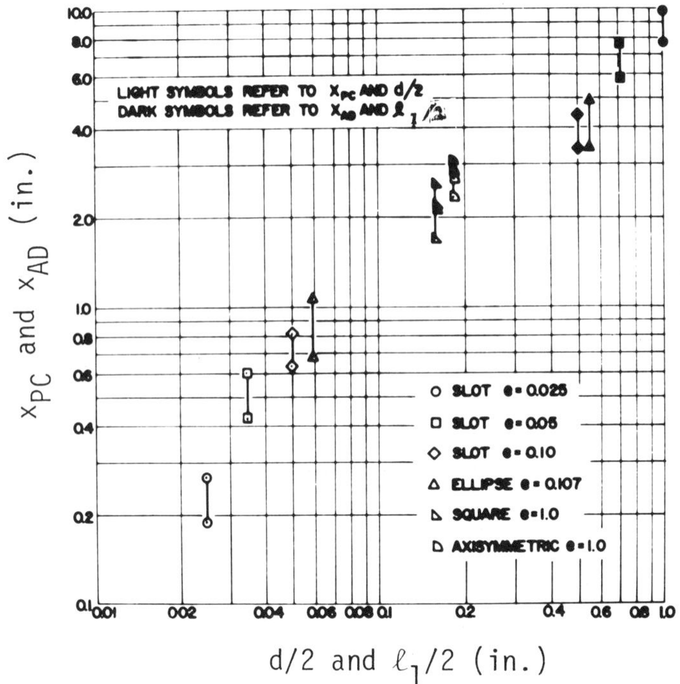

Fig. 129 **Location of x_{PC} and x_{AD} as functions of orifice half-height ($d/2$) and half-width ($l_1/2$).[153]**

3. Hypermixing Nozzles

A "hypermixing" nozzle is a device where a row of small jets is injected at alternating, small positive and negative angles to an essentially coflowing mainstream as shown in Fig. 133 from Ref. 156. The alternating upward and downward jet velocity components interact with each other to produce streamwise vortices that increase mixing. Some measured velocity profiles at a mid-element station and also along the boundary between two elements are given in Fig. 134. Also shown are the predictions of a two-equation model that will be discussed below. Note that the profiles between the individual elements have two peaks that persist quite far downstream.

C. Analysis

1. Mean-Flow Models

In Ref. 151, the Prandtl eddy viscosity model, Eq. (61) is extended on an ad hoc basis to three-dimensional cases as

$$\nu_T = 0.037\,\rho \Psi U_c \tag{167}$$

where

$$\Psi \equiv [y^a_{1/2} z^a_{1/2} (y^a_{1/2} + z^a_{1/2})]^{1/a} \tag{168}$$

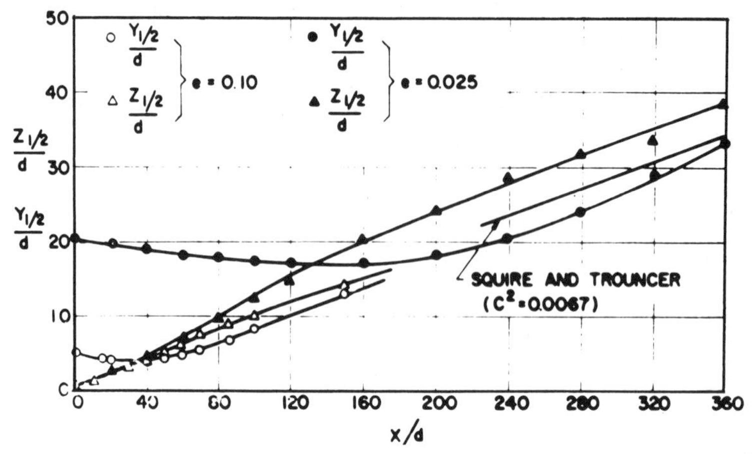

Fig. 130 Streamwise variation of half-height and half-width for two rectangular jets.[153]

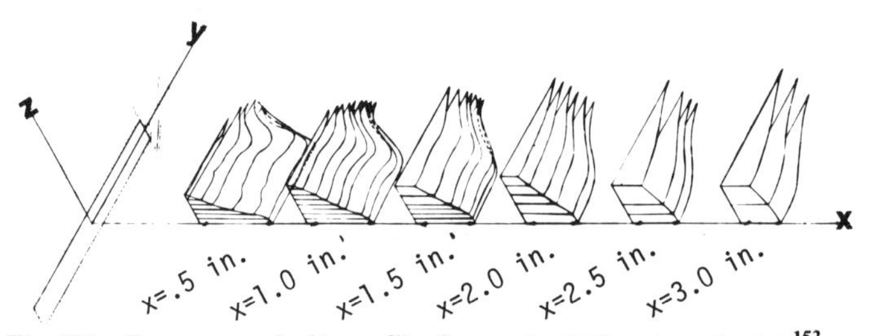

Fig. 131 Transverse velocity profiles for an $e^* = 0.10$ rectangular jet.[153]

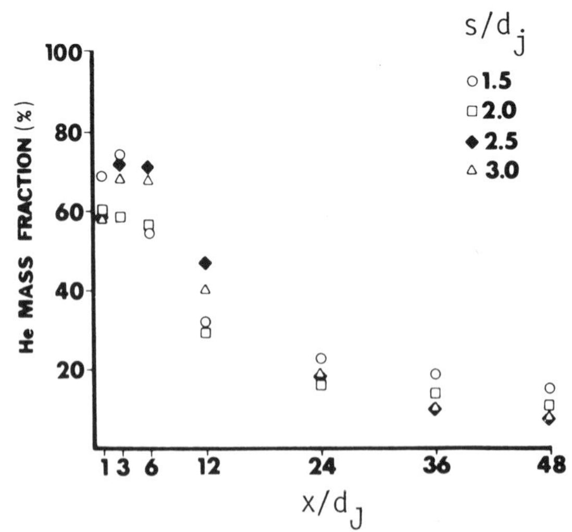

Fig. 132 Streamwise growth of velocity half-radius in particle-laden axisymmetric jets[145]: 1(all gas), 2(32, 0.3), 3(32, 0.56), 4(32, 0.77), 6(72, 0.3), 9(17, 0.3), 10(80, 0.3) (published with permission).

and $(1/a)=0.53$. The boundary-layer form of the three-dimensional equations of motion was also heuristically simplified by dropping the second and third terms in the convective derivative in each equation. The resulting system of equations was solved by a finite-difference method. The predicted variation of the centerline velocity for a 10:1 rectangular jet is compared to the data in Fig. 135, where rather good agreement can be seen.

The work of Ref. 155 uses approximate single-jet solutions to obtain an approximate solution to the adjacent jet problem. The analysis begins with the equation for axisymmetric flow written in Von Mises coordinates (x, ψ):

$$\frac{\partial U}{\partial x}=\frac{1}{U_\infty \psi}\frac{\partial}{\partial \psi}\left[\frac{\rho^2 \nu_T r^2 U}{\rho_\infty^2 U_\infty \psi^2}\psi\frac{\partial U}{\partial \psi}\right] \tag{169}$$

which is linearized with

$$\frac{\rho^2 \nu_T r^2 U}{\rho_\infty^2 U_\infty \psi^2}\approx\frac{\rho \upsilon_T}{\rho_\infty}A(x) \tag{170}$$

where $A(x)$ is a "stretching" factor. With a new streamwise variable $\xi(x)$,

$$\xi\equiv\int_0^x \frac{\rho\nu_T A(x)}{\psi_j \rho_\infty U_\infty}\,\mathrm{d}x \tag{171}$$

we get

$$\frac{\partial U}{\partial \xi}=\frac{\psi_j}{\psi}\frac{\partial}{\partial \psi}\left(\psi\frac{\partial U}{\partial \psi}\right) \tag{172}$$

It is important to note that, since this equation is linear, the solution for the three-dimensional flow produced by multiple jets can be found from the axisymmetric solution for a single jet by simple algebraic superposition. The eddy viscosity was taken from Ref. 77, and $A(x)=\text{const}$ was found by comparison of single-jet predictions and experiment. The predictions for the centerline concentration for the experimental cases of Ref. 155 are shown in Fig. 136. It can be seen that injectant concentration depends on s/d_j up to 2.5 both for theory and experiment; the change of concentration as a function of s/d_j is similar for experiment and analysis; and agreement between theory and experiment is good for the absolute values of the concentration.

2. *Two-Equation Models*

There are currently no one-equation or Reynolds stress models in the literature for three-dimensional jets. The only "higher-order" models available are of the two-equation type.

The k-ϵ model was applied to the three-dimensional nozzle experiments of Refs. 151-154 and in Ref. 157. It was found necessary also to relate C_{ϵ_1} to $(\mathrm{d}U_c/\mathrm{d}x)$, in addition to Eqs. (85) and (86). The relation used is

$$C_{\epsilon 1}=1.14-5.31\frac{y_{1/2}}{U_c}\frac{\mathrm{d}U_c}{\mathrm{d}x} \tag{173}$$

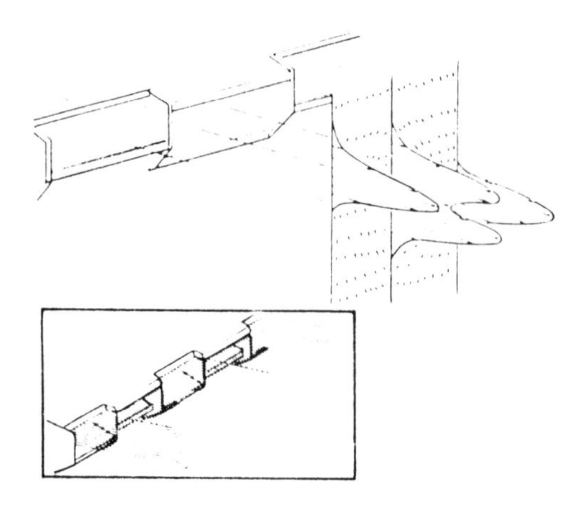

Fig. 133 Schematic of a hypermixing nozzle.[156]

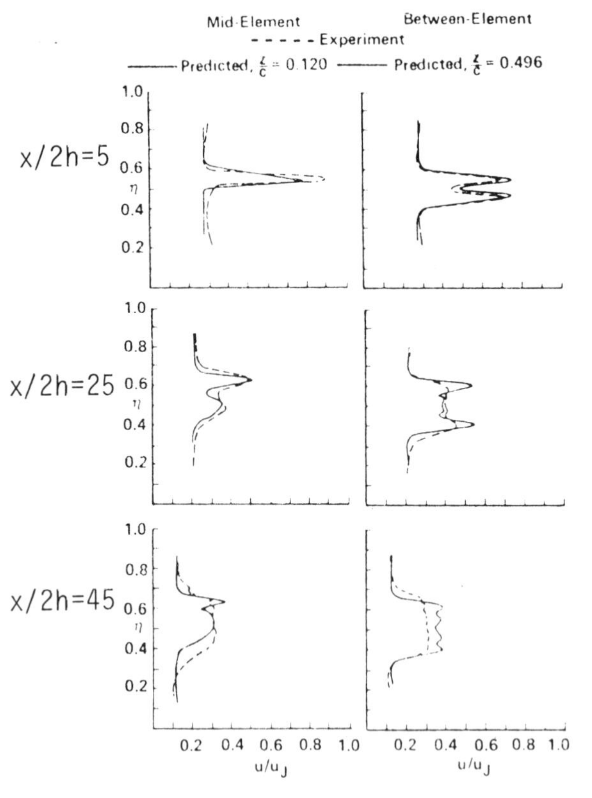

Fig. 134 Comparison of prediction with a two-equation model and experiment for a hypermixing nozzle flowfield.[156]

This equation has been developed only for and is suggested only for jets in stagnant surroundings. The full, boundary-layer form of the three-dimensional equations was solved numerically.

Prediction of the centerline velocity variation for a 10:1 rectangular jet is compared with data in Fig. 137, and a comparison of half-width growths is

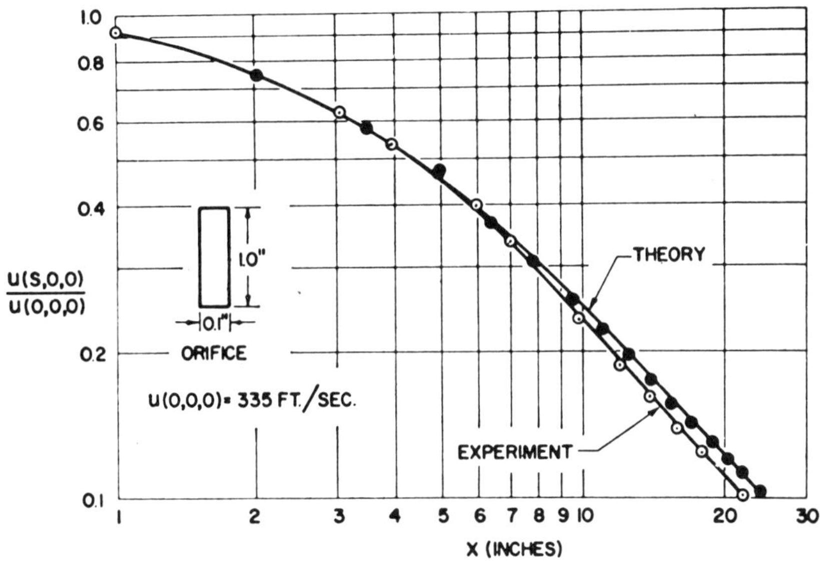

Fig. 135 Comparison of prediction with a mean-flow model and experiment for a rectangular jet.[151]

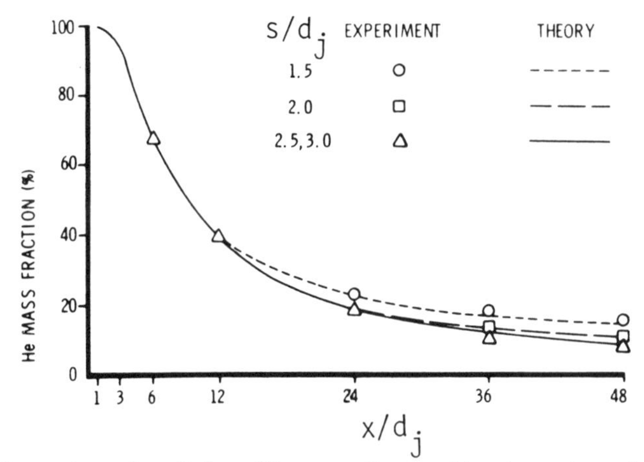

Fig. 136 Comparison of prediction with a mean-flow model and experiment for four adjacent helium jets in a $M=4$ airstream[155]

given in Fig. 138. The agreement is generally very good, although the growth of $z_{1/2}$ is slightly underpredicted. It was necessary to make estimates of the initial lateral velocity distributions in the jet to obtain these good results. The "saddles" in the velocity profiles observed experimentally (see Fig. 131) are not, however, predicted by this analysis.

In addition to experiments, Ref. 156 also reported the application of a k-ϵ model to the hypermixing nozzle problem. The unmodified $k\epsilon 1$ model of Ref. 81 with the constants suggested for planar flows was applied directly. The full

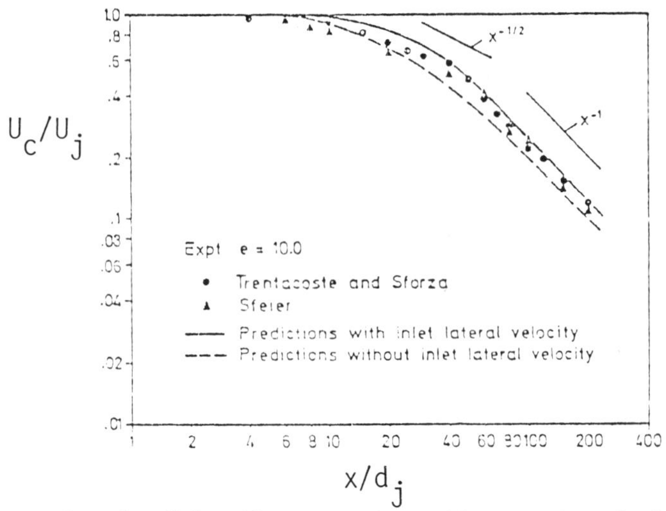

Fig. 137 Comparison of prediction with a two-equation model and experiment for the centerline velocity variation of a 10:1 jet[157] (published with permission).

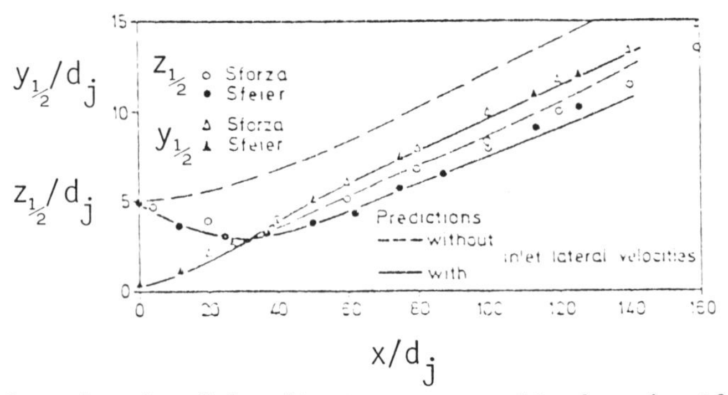

Fig. 138 Comparison of prediction with a two-equation model and experiment for half-width growth of a 10:1 jet[157] (published with permission).

boundary-layer equations were solved numerically. The results of this analysis are compared with the data in Fig. 134. Quite good agreement was achieved for the mid-element profiles. The results for the between-element profiles must be judged as good in light of the complexity of the flow in that region.

D. Discussion

The reader has perhaps noticed that no turbulence data for these three-dimensional flows have been reported. The lack of this type of information not only is regrettable in a general way, but it significantly retards the development of suitable turbulence models.

VIII. TRANSVERSE INJECTION

A. Problem Definition

In this chapter, we shall be concerned with jet injection at large (order of radians) angles to a moving mainstream. This, of course, produces strongly three-dimensional flowfields, but of a type generally different from those treated in the last chapter. This class of flows is often encountered in practice; smokestacks, some fuel injection systems, and sewage and cooling water outfalls can be cited as a few representative examples. One also can have two-phase cases, but the main emphasis here is on single-phase cases, since much less basic data are available for the two-phase situation. Finally, one or both fluid streams may be supersonic. For the essentially coaxial cases discussed to this point, that is not an important matter; however for transverse injection, supersonic flows require special treatment. Such cases are discussed in this chapter in separate sections. Lastly, cases where buoyancy forces are important are reserved for the next chapter.

B. Experimental Information

1. Low-Speed, Single-Phase Flows

The characteristics of the flow for a case with 90-deg injection can be seen in Figs. 139 and 140 from Ref. 5. The "kidney-shaped" nature of the jet as it is deflected and distorted by the cross stream is particularly noteworthy. The simplest quantity of engineering interest is the gross penetration of the mass of injected fluid into the mainstream. The trajectory of the center of the jet plume as a function of velocity ratio U_j/U_∞, and injection angle is shown in Figs. 141 and 142 from Ref. 158 using data from Refs. 5 and 159-161. The "theory" curves will be discussed later. For 90-deg injection, trajectory results can be correlated as in Fig. 143 from Ref. 162. Here h is the vertical penetration, and ξ is the arc length along the jet centerline trajectory (see Fig. 144). The next quantity of interest is the growth of the width of the mixing zone in the plane containing the trajectory and the direction perpendicular to that plane. These widths are denoted here as $\Delta\zeta$ and Δz, and some results are reproduced as Figs. 145 and 146.

An interesting test series is presented in Ref. 163, where a transverse jet was injected from a symmetrical wing section (NACA 0021) at various angles of attack (and thus lift coefficients C_L). The results showed an increase in penetration over injection through a flat (nonlifting) surface at the same $\bar{q}$ as a function of C_L as shown in Fig. 147.

Turning now to some details of the flow, we show the decay of the maximum velocity in the jet cross section in Fig. 148 from Ref. 160. Free jet results are also shown for comparison. It is not surprising that the transverse jet mixes faster, probably a direct result of the vortices induced in the jet

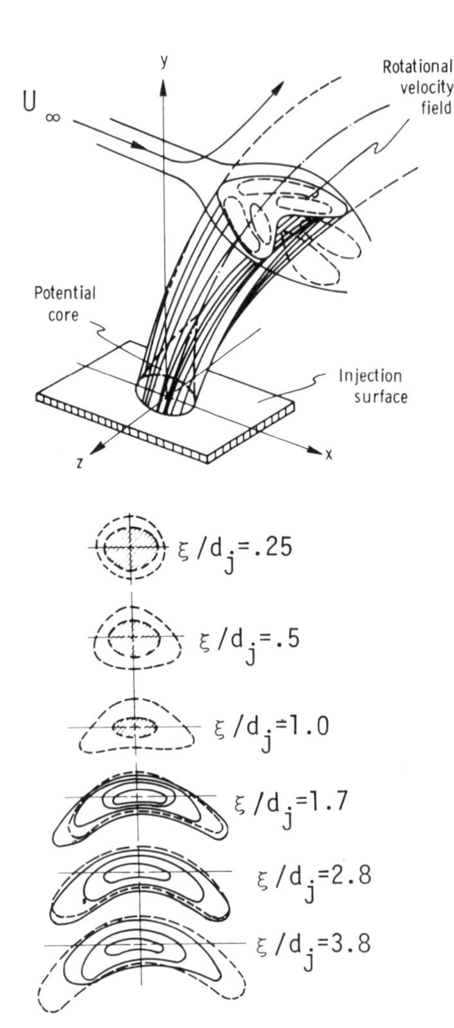

Fig. 139 Schematic of transverse injection flowfield[5] (published with permission).

Fig. 140 Cross-sectional pressure contours in a transverse jet with $U_j/U_\infty = 2.2$[5]; solid and dashed lines correspond to constant total and static pressure, and the shaded areas denote the potential core (published with permission).

plume. Velocity profiles in the lateral direction are shown in Fig. 149, where a near-similarity condition can be observed. Isovels in the plane of the trajectory are shown in Fig. 150 from Ref. 164. Entrainment into the jet plume was measured and reported in Refs. 160 and 165. Figure 151 shows some results from Ref. 160.

In Ref. 166, velocity measurements in the jet are used to infer the properties of the vortices in the plume via back-calculation. The strength of the vortices as a function of injection angle, velocity ratio, and downstream distance is shown here in Fig. 152. It can be seen that the influence of injection angle is significant only for the larger velocity ratio. Information on vortex core size and spacing is also given in Ref. 166.

The variation of the axial turbulence intensity along the jet centerline for three velocity ratios is shown in Fig. 153 from Ref. 160. Also shown for comparison are some coaxial jet results from Ref. 40. The turbulence is higher

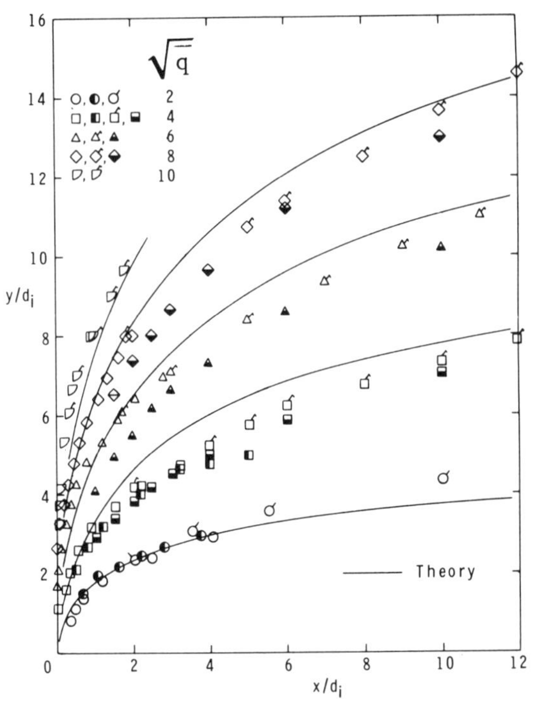

Fig. 141 Trajectories for transverse, $\theta = 90$ deg, air jets into air for various values of the parameter $\bar{q} = \rho_j U_j^2 / \rho_\infty U_\infty^2$.[158]

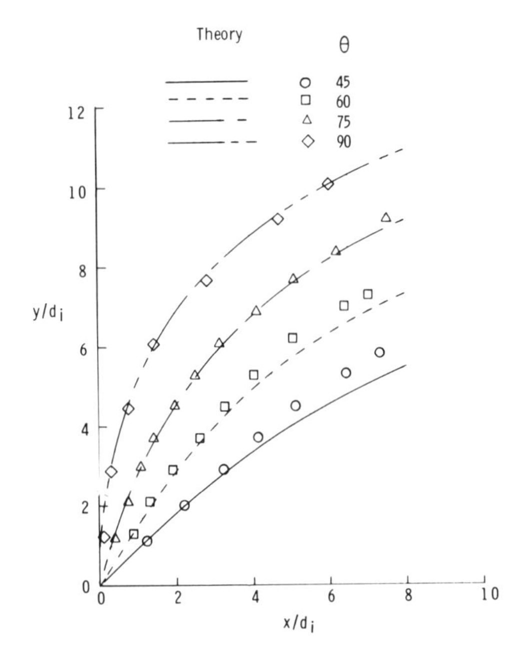

Fig. 142 Trajectories of transverse jets for various injection angles with $(\bar{q})^{1/2} = 6.3$.[158]

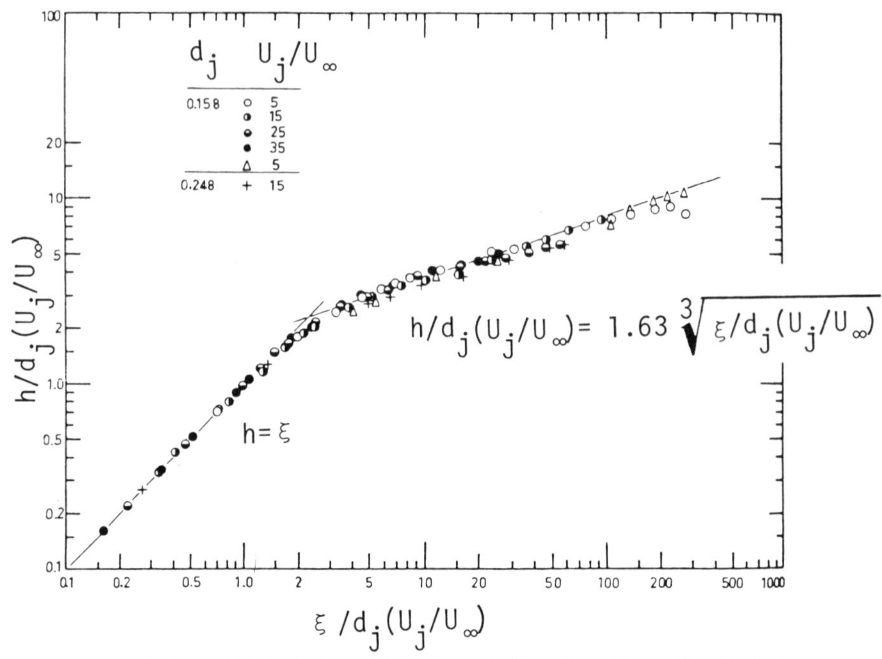

Fig. 143 Correlation of the trajectory of the jet centerline along the arc length for transverse $\theta = 90$ deg, injection[162] (published with permission).

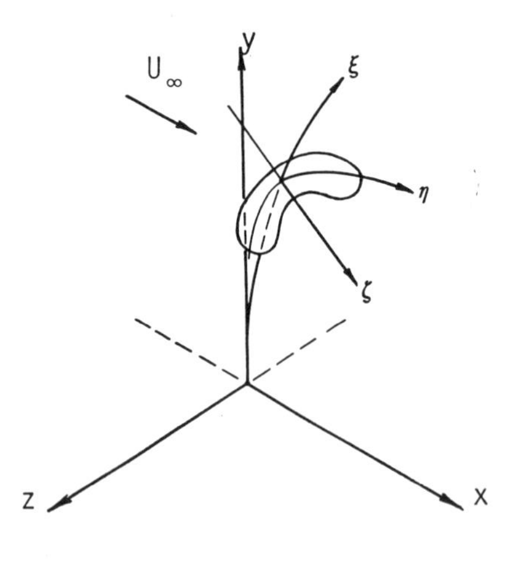

Fig. 144 Coordinate system for transverse jets.

in the transverse jet case, which is consistent with the more rapid mixing of the mean flow noted earlier.

2. *Transverse, Particle-Laden Jets*

The situation of a particle-laden fluid jet injected transverse to a fluid stream is of interest because of the possible inertial effects of the particles. One

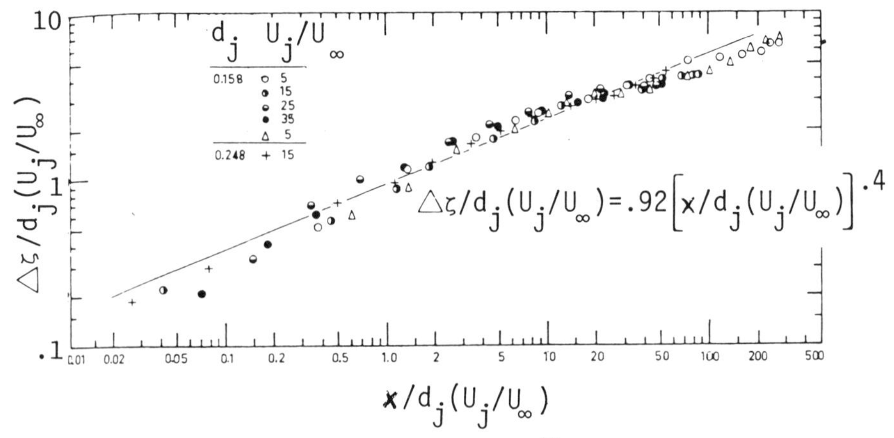

Fig. 145 Spread in the ζ direction of a transverse jet[162] (published with permission).

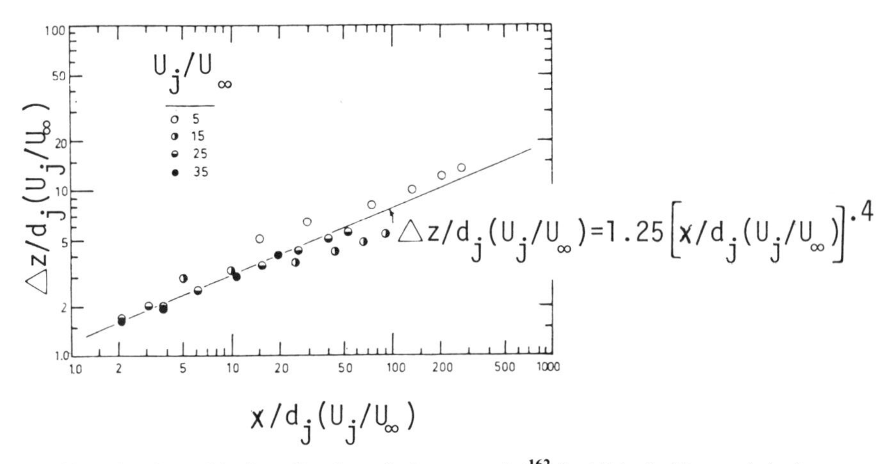

Fig. 146 Spread in the z direction of a transverse jet[162] (published with permission).

might anticipate an appreciable "separation" of the injectant fluid and particle streams under the influence of the cross flow, at least for large particles.

The range of the variables that has been studied is very limited, and it is difficult at this time to arrive at firm general conclusions. In Refs. 167 and 168, gas jets loaded with small particles (e.g., 15-μ silicate in Ref. 167) were injected at 90 deg to low-speed airstreams. The primary result is the trajectory of the center of mass of the solid phase under various conditions. Although the scatter in the data is large, the results of Ref. 167 are correlated reasonably well (see Fig. 154) by the equation

$$y/d_j = 1.92\,(\bar{q})^{0.335}\,(x/d_j)^{0.33} \tag{174}$$

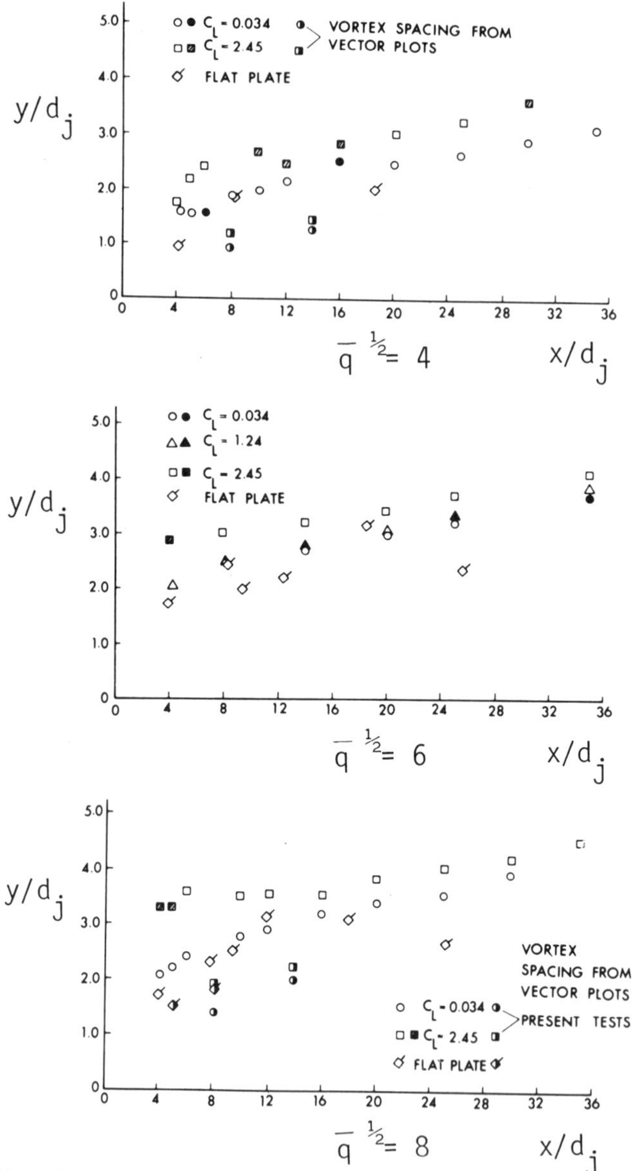

Fig. 147 Effect of lift coefficient on penetration for transverse injection from an NACA 0021 airfoil.[163]

The data of Ref. 168 indicate about 15% greater penetration. Both of these sets of data show a greater penetration (by roughly 20-30%) than single-phase cases at the same $\bar{q}$.

A different physical arrangement was studied in Ref. 169, as shown in Fig. 155. The outer stream is directed toward the centerline at various angles. The

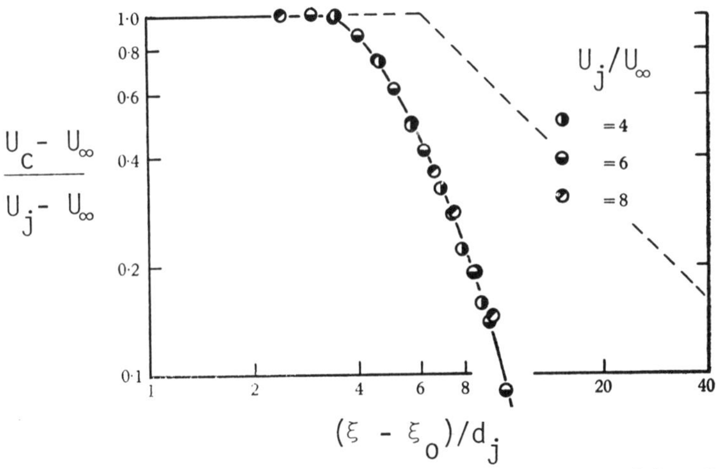

Fig. 148 Variation of centerline velocity along the jet trajectory measured from the virtual origin[160] : -------free jet data (published with permission).

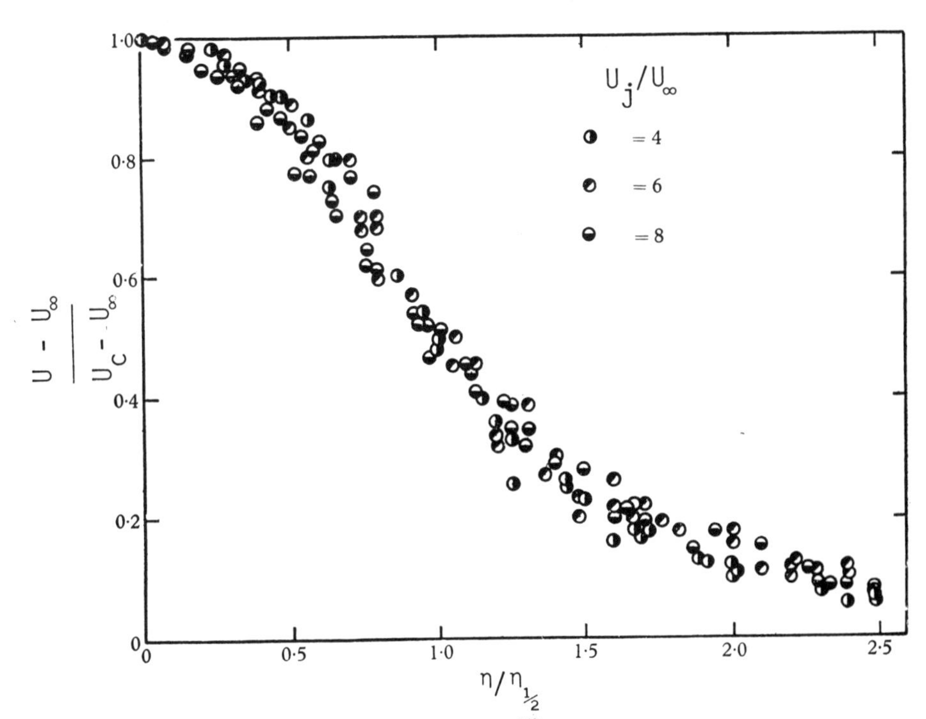

Fig. 149 Transverse profiles across the jet[160] (published with permission).

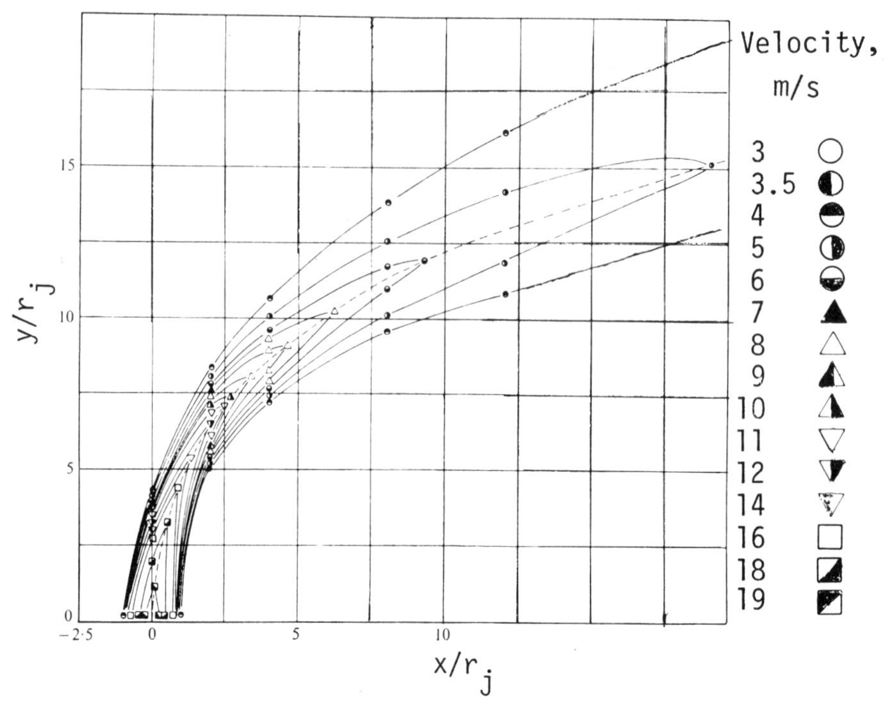

Fig. 150 Velocity map in a transverse jet with $U_j/U = 3.95$ and $U_j = 16.95$ m/s[164] (published with permission).

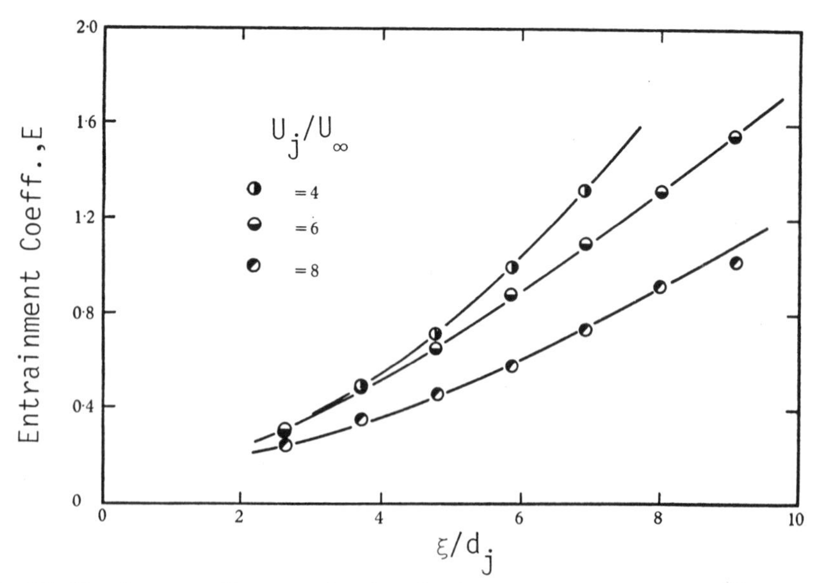

Fig. 151 Entrainment coefficient as a function of arc length along the jet trajectory[160] (published with permission).

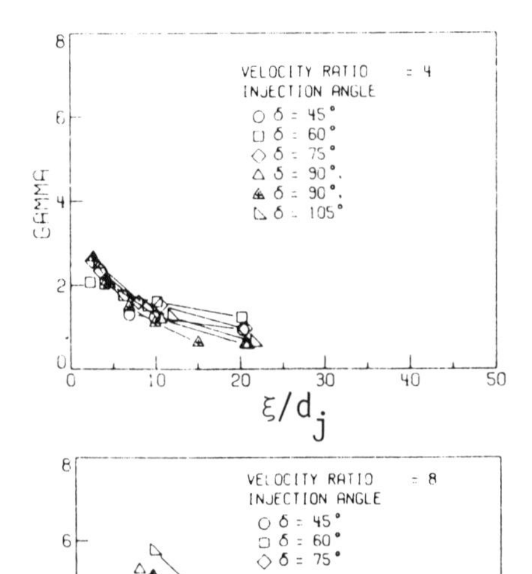

Fig. 152 Strength of the vortices induced in transverse jets.[166]

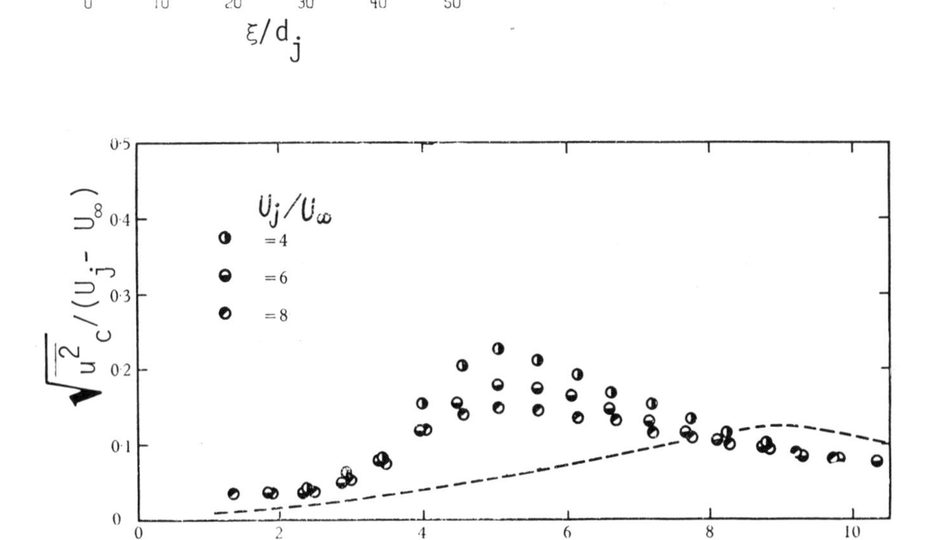

Fig. 153 Turbulence intensity along the trajectory of a transverse jet[160]: -----free jet data (published with permission).

tests to be discussed here were conducted with 6-μ aluminum powder at 20% by weight in a helium (10 mole %)-air jet. The effect of nonparallel injection was large on the axial decay of the centerline concentration of both the solid phase and the "foreign" gas (helium) in the primary jet, as can be seen in Figs. 156 and 157. It is interesting to observe that the effect of injection angle on the gas mixing rate is a relatively smooth function of angle from 0 to 90 deg, whereas a rather precipitous increase in mixing rate occurs in going from 0 to 30 deg for the solid phase. Further increases in angle have only a minor effect on the particle mixing rate.

3. *Transverse Jets into Supersonic Flows: Gaseous Jets*

Transverse injection into supersonic streams is of engineering interest in several applications, including, as examples, thrust vector control (TVC), fuel injection, and thermal protection. Since such a jet presents an "obstruction"

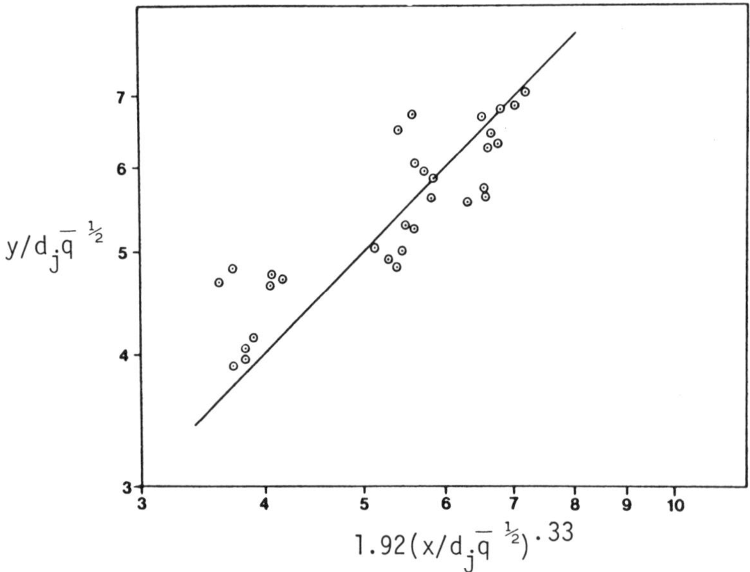

Fig. 154 Correlation of plume centerline trajectory for a transverse particle-laden jet[167] (published with permission).

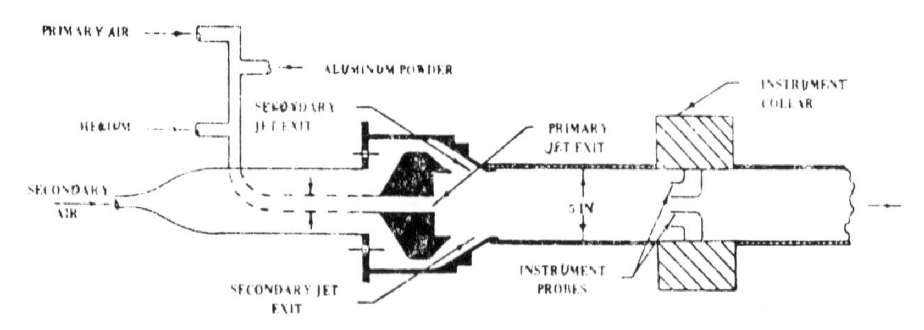

Fig. 155 Schematic of the flow arrangement studied in Ref. 169.

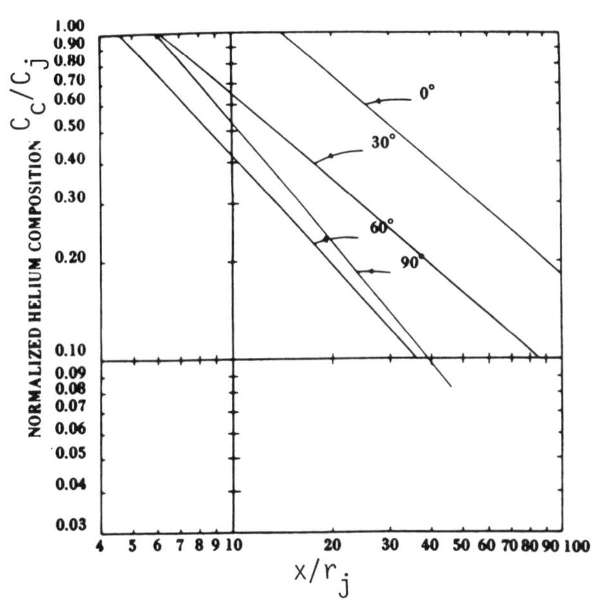

Fig. 156 Centerline helium concentration vs streamwise distance for an air-helium-particle jet into a surrounding airstream at various angles.[169]

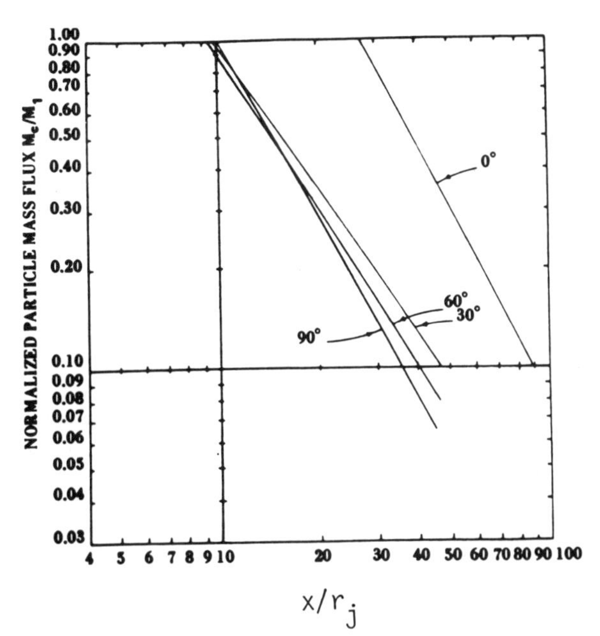

Fig. 157 Centerline particle mass flux vs streamwise distance for an air-helium-particle jet into a surrounding airstream at various angles.[169]

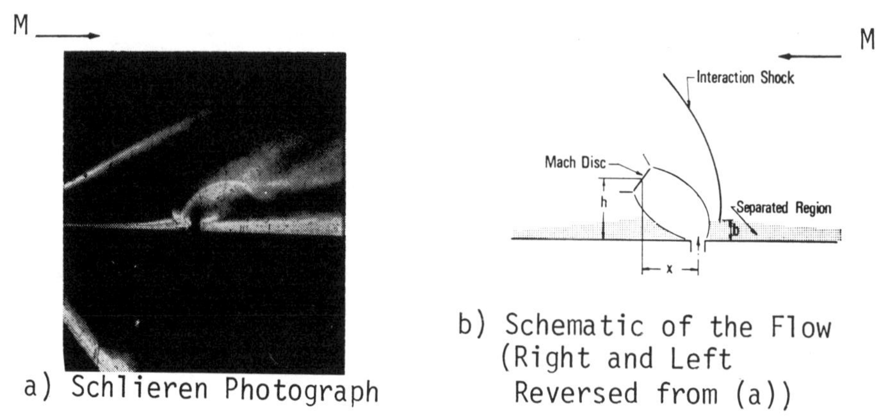

Fig. 158 Flowfield observed for an underexpanded transverse gas jet into a supersonic stream.[171]

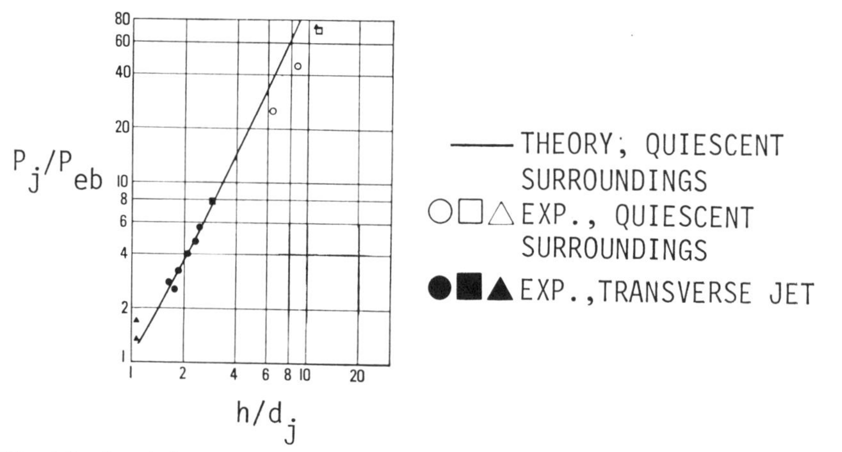

Fig. 159 Correlation of the Mach disk height in an underexpanded transverse gas jet into a supersonic stream using theory and experiment for an underexpanded jet into quiescent surroundings.[171]

to the main supersonic flow, an "interaction shock" is generally produced, and the whole flow differs from low-speed cases in important ways.

One of the earliest studies of these flows was presented in Ref. 170. The emphasis there and in most of the early works was on the pressure distribution and the side force produced on the adjacent surface for the TVC application, although some details of the jet and interaction shock were given. In Ref. 171, the case of a sonic, underexpanded transverse jet was considered, and the resulting flowfield shown in Fig. 158 was related to that for an underexpanded jet into quiescent surroundings. This relationship was accomplished through the notion of an "effective" back pressure P_{eb}. It was then possible to use the experimental and theoretical results for the quiescent surroundings case to correlate data for the height (penetration) of the Mach disk in the transverse

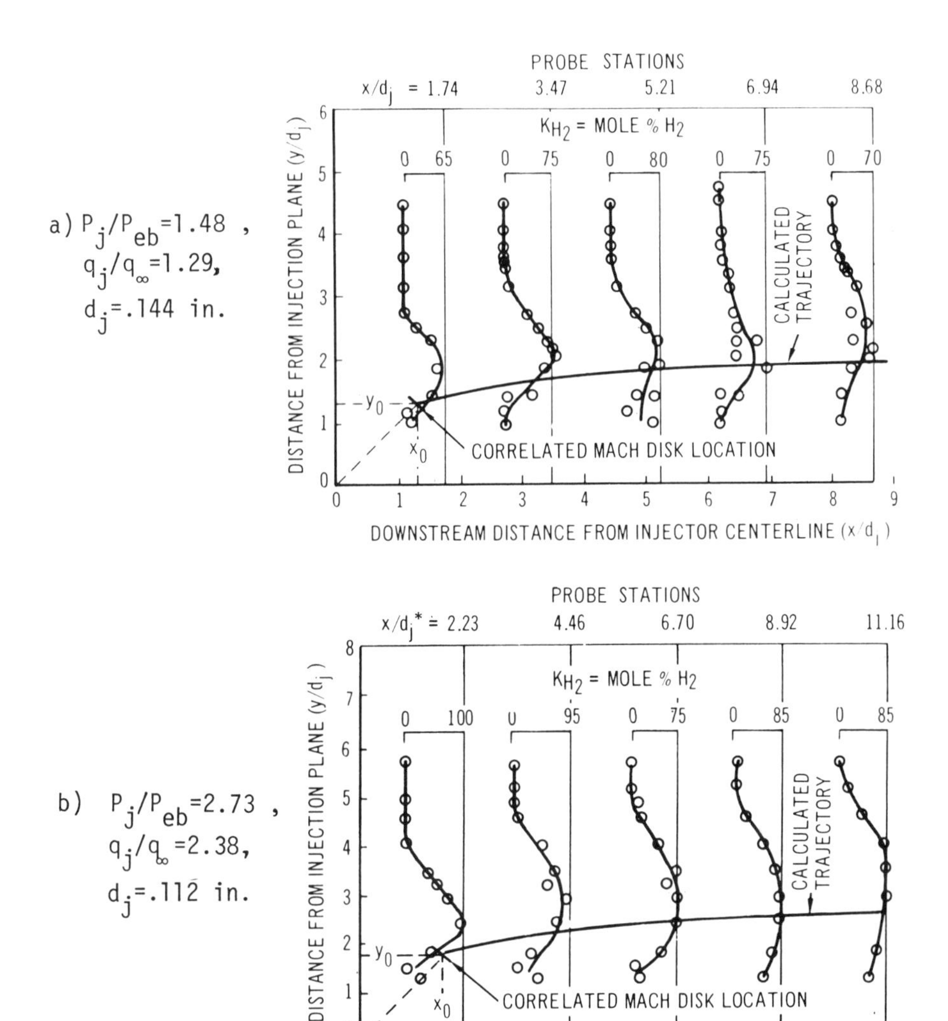

Fig. 160 Comparison of prediction and experiment for an underexpanded transverse H_2 jet into $M_\infty = 2.7$ air.[173]

jet flow. This is shown in Fig. 159 using $P_{eb} = 0.8\ P_2$, where P_2 is the static pressure behind a normal shock in the main flow. Similar results for supersonic, underexpanded transverse jets are given in Ref. 172. The effect of the shape of injectors with the same cross-sectional area on the height of the Mach disk was studied in Ref. 173 and found to be unimportant. From a practical viewpoint, these results together show that it is impractical to try to increase transverse gaseous penetration substantially into a supersonic crossflow by increasing injection pressure markedly, varying the shape of the injector, or employing supersonic injection.

In addition to the initial penetration of the transverse jet as given here in terms of the height of the Mach disk, the subsequent trajectory of the jet and mixing along that trajectory are also of interest. Some results for underexpanded, sonic, H_2 injection with $M=2.7$ from Ref. 173 are plotted in Fig. 160. Concentration profiles across the jet plume are given in Fig. 161, where it can be seen that the effects of injector shape are again small.

The shape of the interaction shock is important in some applications, and some results from Ref. 174 are shown here in Fig. 162. The results of a simple

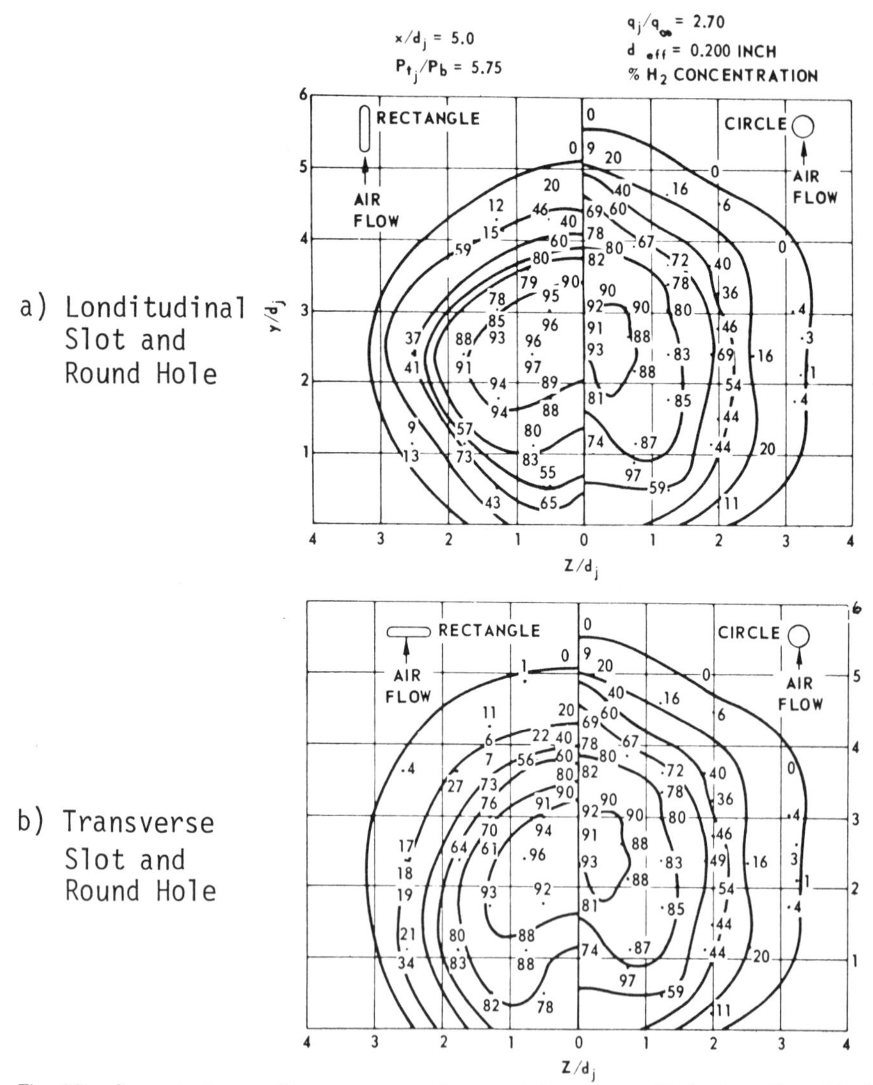

Fig. 161 Concentration profiles across an underexpanded transverse H_2 jet into $M_\infty = 2.7$ air with different injector shapes.[173]

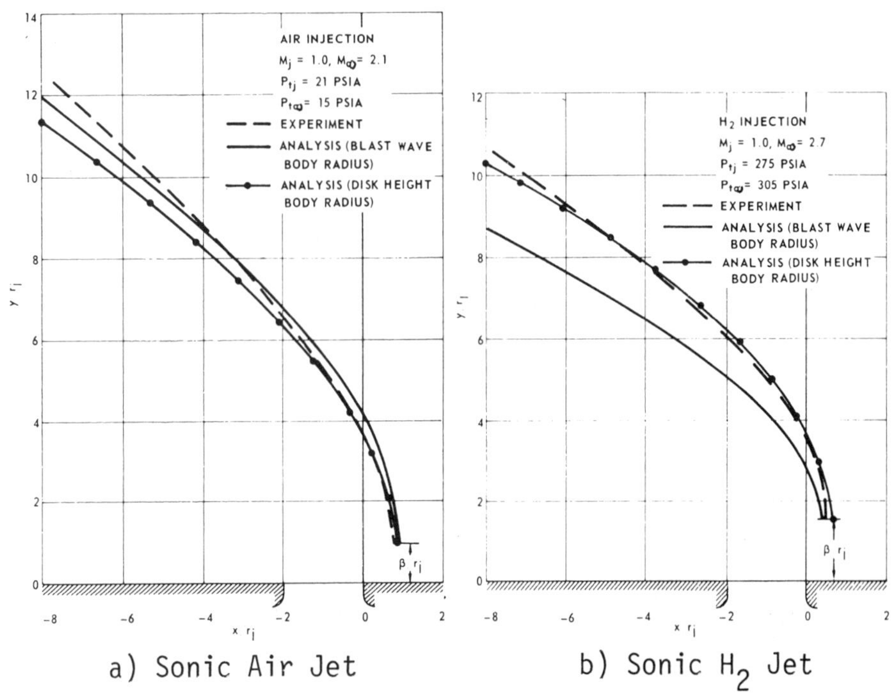

Fig. 162 Comparison of predictions and experiment for the interaction shock shape of transverse gas jets into supersonic flow.[174]

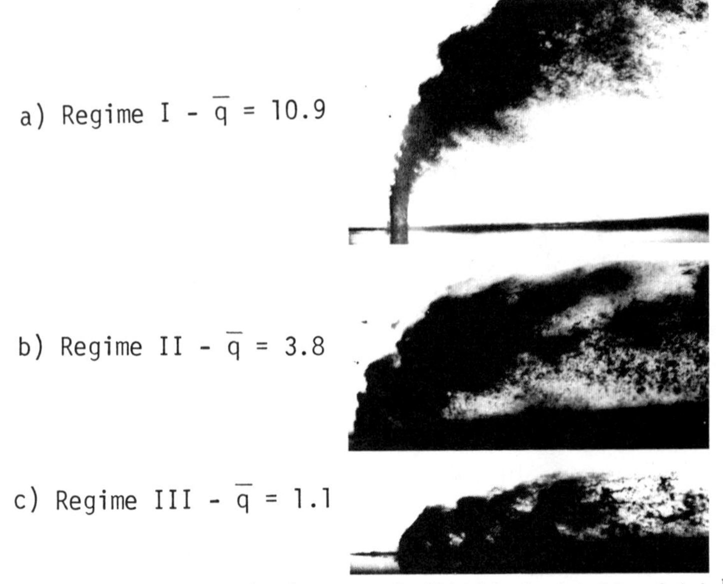

Fig. 163 Spark photographs of transverse liquid jet injection into $M_\infty = 2.4$ air.[178]

Fig. 164 Nanophoto (15×10^{-8} s) of transverse liquid jet injection into $M_\infty = 3.0$ air.

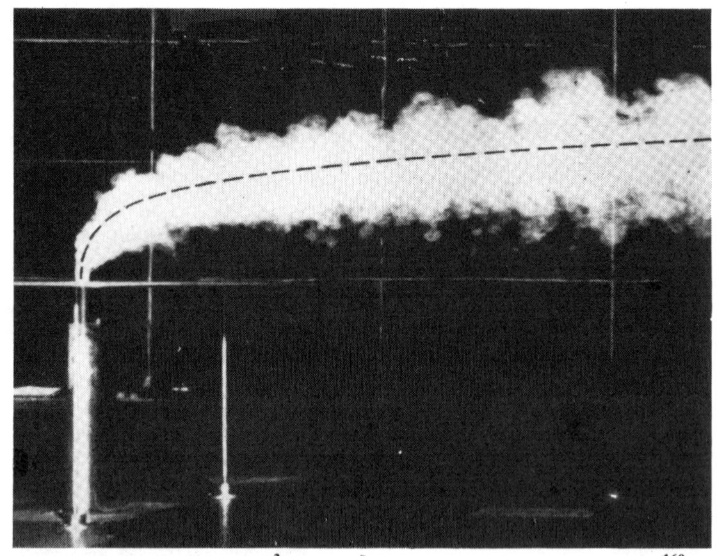

Fig. 165 Direct photograph (0.8×10^{-3} s) of a transverse gas jet in low-speed flow[160] (published with permission).

"analysis" based upon an equivalent solid body obstruction are also shown, along with the less successful "blast-wave" model.

4. *Transverse Jets into Supersonic Flow: Liquid Jets*

This, of course, represents a two-phase flow, but the matters of primary interest here will be the gross penetration and behavior of the jet, not the

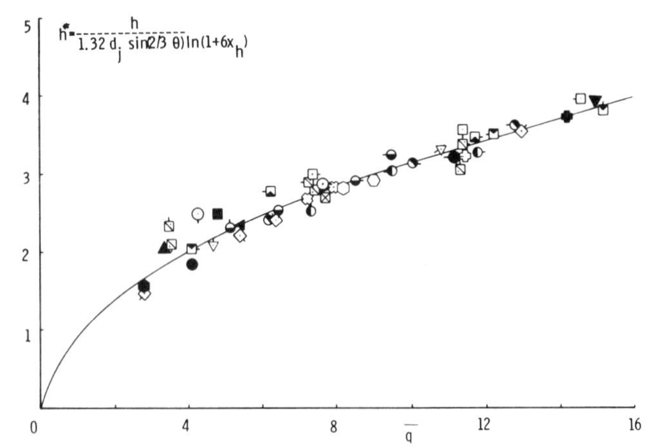

Fig. 166 Correlation of liquid jet penetration into supersonic flow.[181]

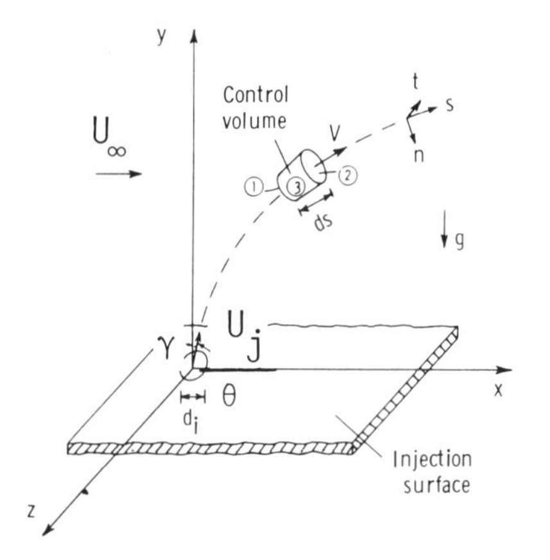

Fig. 167 Flowfield model for transverse jet trajectory analysis.

droplet processes that were discussed in Chapter VI. There have been a number of works aimed at these gross features of the flow, e.g., Refs. 175-181. The interaction of the liquid jet column with the main supersonic stream is highly unsteady, and only high-speed (~20,000 pictures/s) motion pictures are really adequate to display all of the features of the flow. Some stop-action (10^{-6} s) stills are shown in Fig. 163 from Ref. 178 to show how the flowfield develops for various ranges of the important parameter $\bar{q}$. A shorter-duration photograph (10^{-8} s) reveals even more details, as in Fig. 164. For the higher $\bar{q}$ range, such stills and high-speed motion pictures show an astonishing similarity to the presumably unrelated case of smokestack plumes. A photograph of that type of flow from Ref. 160 is shown here for comparison in Fig. 165.

The wave processes on the jet are clearly very important for the processes of breakup and atomization. Careful studies have shown that the waves are not produced by any fluctuations in the initial jet stream and, perhaps surprisingly, that they are not substantially influenced by the physical properties (viscosity and surface tension) of the injectant.[177] Thus, these waves appear not to be of the simple character of those that are well known on jets into quiescent surroundings. Perhaps the apparently superficial relation to smokestack flows may, in reality, be important.

Different workers have developed correlation formulas for the gross penetration of the liquid plume into the cross flow, including the effects of various parameters. Taking the influence of $\bar{q}$ from Ref. 178, x/d_j from Ref. 180, the aspect ratio d_f/d_s of the injector from Ref. 179, and injection angle θ from new work, an overall correlation was developed in Ref. 181. The complete formula is

$$\frac{h}{d_j} = 1.32(\bar{q})^{1/2} C_d \left(\frac{d_{eq}}{d_f}\right)^2 \left(\frac{d_f}{d_s}\right)^{0.46} \ln\left[1+6\left(\frac{x}{d_j}\right)\right] \sin\left(\frac{2\theta}{3}\right) \tag{175}$$

The adequacy of the correlation is indicated in Fig. 166 for a number of varied experimental cases.

It should be noted that the liquid jet penetration results do show useful increases in penetration with a suitable change in injector shape and/or increased injection pressure. This is in contrast to gaseous jet results where the formation of the Mach disk in the jet flow plays a dominant role. Lastly, penetration results for liquid injection into a high-speed, but subsonic, cross flow are available in Ref. 182.

C. Analysis

1. *Trajectory Analyses*

In this class of analysis, one is concerned only with predicting the penetration and spread of the jet plume. The details of the flow such as

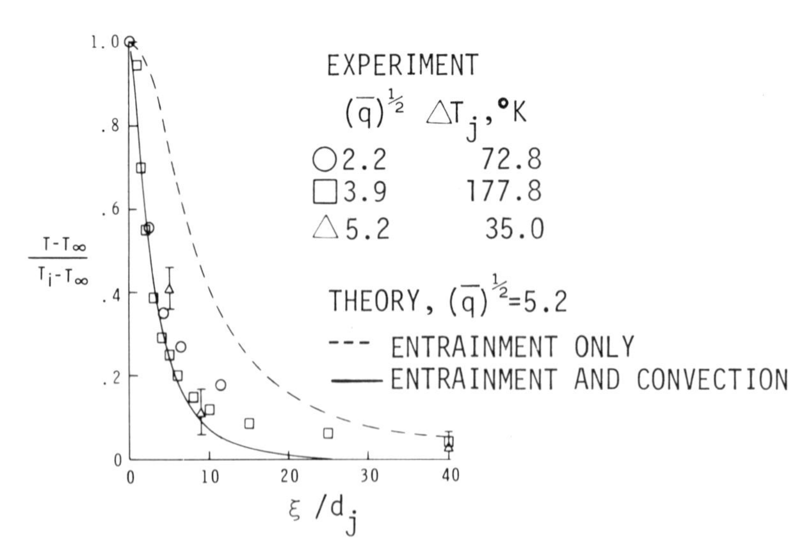

Fig. 168 Comparison of prediction and experiment for temperature variation along the jet trajectory for transverse injection of a heated jet.[158]

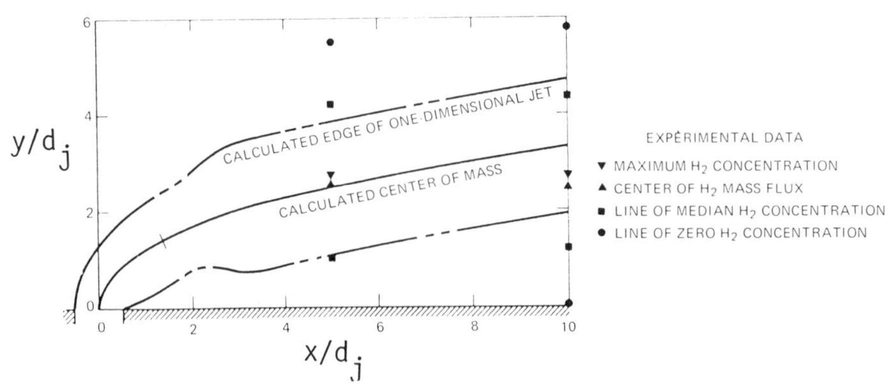

Fig. 169 Comparison of prediction with the jet penetration analysis of Ref. 184 and experiment for a transverse H_2 jet into $M_\infty = 2.7$ air with $P_{0j}/P_{0\infty} = 1.6$.

velocity profiles are not treated directly. A schematic of a model of this type is given here as Fig. 167. Looking at an element of the jet column, the various overall forces are modeled, and the trajectory of the column may be predicted. One of the earliest such analyses was developed in Ref. 5; however, the variation of normal momentum was neglected, and it proved necessary to use unrealistically high values of the drag coefficient acting on the jet column to obtain satisfactory results. That assumption was relaxed in Ref. 183, and further refinements were introduced in Ref. 158. The "theory" curves on Figs. 141 and 142 are the result, and the rather good agreement with experiment can be noted.

It is a simple matter to add an energy equation to this type of analysis and thus treat heated or cooled jet cases. This was done in Ref. 158, and the results are compared with experiment in Fig. 168. The effects of a nonuniform crossflow can also be easily incorporated as in Ref. 158, but no detailed experiments were available for comparison.

For the supersonic main flow cases, the only additional information required is suitable drag coefficient data at the normal Mach numbers of interest. If the jet flow is at least sonic and underexpanded, then the trajectory analysis is begun at the Mach disk location. The calculated trajectories on Fig. 160 were obtained in that way using the analysis of Ref. 183. It can be observed that there is little further penetration after the Mach disk. This is a result of the total pressure loss through the Mach disk (shock). The work was carried somewhat further in Ref. 184, from which Fig. 169 is taken. Quite good predictions of the gross behavior of the jet are clearly obtainable with this level of analysis. For liquid jets into either low-speed or supersonic crossflows, the breakup of the jet column influences penetration directly, and a simple trajectory analysis of the type discussed here is adequate only until the jet column "fractures."

2. *Differential, Mean-Flow Models*

The differential analysis of these flows must confront a three-dimensional problem where none of the components may reasonably be neglected as small.

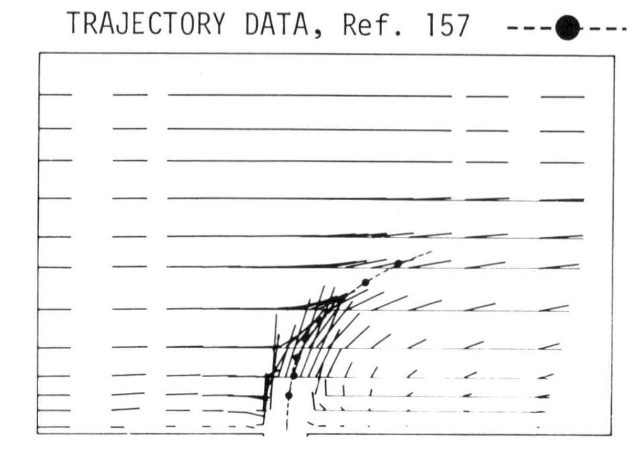

Fig. 170 Comparison of a prediction with a mean-flow model in the Navier-Stokes equations and experiment for the trajectory of a transverse jet with $U_j/U=4$.[186]

Thus, these treatments are generally based upon the full, three-dimensional Navier-Stokes equations, and computing time and storage requirements tend quickly toward the prohibitive.

The treatment of Ref. 185 was based upon an unsteady, differential-integral formulation, and only the early time development of the flow was obtained. A steady, velocity-vorticity-temperature formulation was employed in Ref. 186, and complete solutions were obtained. A comparison of a prediction with data from Ref. 160 is shown in Fig. 170. The line segments on the figure denote the predicted magnitude and direction of the local velocity vectors in the plane of the jet trajectory. This prediction was obtained with a crude turbulence model. An eddy viscosity was taken as constant at the initial value that would be predicted by the Prandtl coaxial jet model:

$$\nu_T = 0.025(U_j - U_\infty)d_j \tag{176}$$

It was found that the predicted trajectory of the jet was not very sensitive to the value of the eddy viscosity by varying the value of the proportionality constant. Presumably, inviscid effects are dominant for that gross quantity.

3. *Higher-Order Models*

Again in this case, the higher-order turbulence models have not been applied.

IX. BUOYANCY FORCE EFFECTS

A. Introduction

There are a number of practical situations where density differences in a jet mixing flow produce buoyancy forces that are significant in comparison to other fluid forces. The ratio of buoyant to viscous forces is expressed in terms of a Grashof number, which for a jet is written as

$$Gr \equiv \frac{g(\rho_\infty - \rho_j)d_j^3}{\rho_j \nu^2} \tag{177}$$

The ratio of inertia to buoyant forces is expressed as a Froude number:

$$Fr \equiv \frac{U_j^2}{g[(\rho_\infty - \rho_j)/\rho_j]d_j} \tag{178}$$

[Most often, the square root of Eq. (178) is taken as the Froude number. Equation (178) may, however, change sign.] These are related through the jet Reynolds number

$$Re \equiv U_j d_j / \nu \tag{179}$$

as

$$Fr \equiv Re^2 / Gr \tag{180}$$

If there is a density gradient in the surroundings of the jet, an additional parameter appears:

$$S^* \equiv -\frac{g}{\rho_a(0)} \frac{d\rho_a(x)}{dx} \tag{181}$$

where $\rho_a(x)$ is the ambient density distribution.

A comprehensive review of vertical, buoyant jets has recently been published,[187] and so only highlights from Ref. 187 and some more recent information for that case will be included here. Figures 171 and 172 show the types of flows that might be encountered. For the "plume" case, there is no well-defined U_j, and so the characteristic velocity in the various characteristic numbers is generally taken as (ν/d_j).

We shall also be interested here in cases where a buoyant jet is injected at an angle to the vertical (gravity). Cases that also have a crossflow are treated as extensions to the material in Chapter VIII. Finally, some discussion of turbulent wakes mixing in a stratified environment will be included.

B. Experiment

1. Buoyant Jets and Plumes

With a nonstratified ambient fluid, a vertical buoyant jet with initial momentum behaves as a nonbuoyant jet near the exit and as a pure plume in

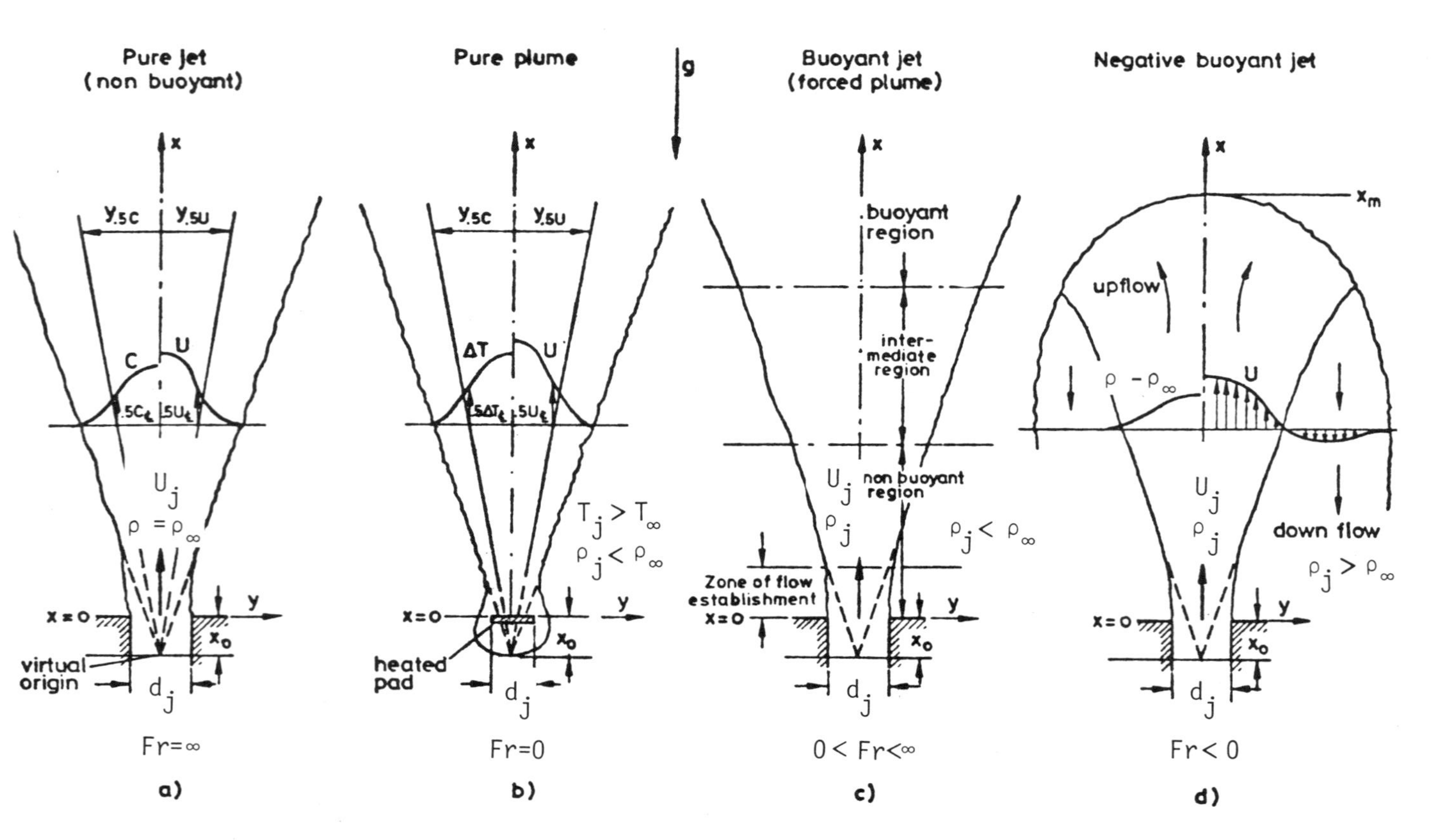

Fig. 171 Schematic of jet and buoyant plume flows[187] (published with permission).

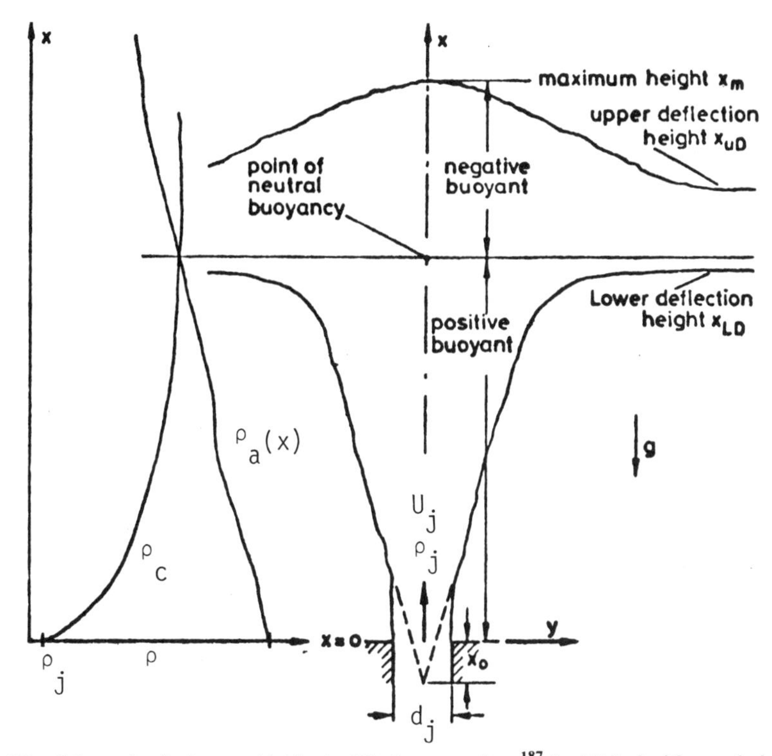

Fig. 172 Schematic of a buoyant jet in stratified surroundings[187] (published with permission).

the far field, even if the Froude number is large. Between these two regimes lies a transition region. Clearly, the boundaries between these regions will depend upon the Froude number. This can all be seen in Figs. 173 and 174.

For a negatively buoyant jet, the experiments of Ref. 195 relate the maximum vertical height to Froude number as shown in Fig. 175.

With linearly stratified surroundings, the results of several investigators[196-201] can be used to obtain the variation of the upper and lower deflection heights with the parameters of the flow as given in Fig. 176. There is relatively little turbulence information for plumes available, and the two experiments in the literature are not in agreement. A laser Doppler velocimeter and a thermistor were applied to a plane plume in Ref. 189, whereas a hot-wire anemometer was used for a round plume in Ref. 202. The methods of Ref. 189 produced results in nonbuoyant jets that were much too high,[187] and so they must be regarded as suspect here also. Both sets of data are plotted in Fig. 177. Restricting attention to the results of Ref. 202, it can be seen that $\sqrt{\overline{T'^2}}/\Delta T_c$ and $\overline{uT'}/U_c T_c$ are higher than in a nonbuoyant jet.

Some high Froude number experiments with heated, vertical air jets are reported in Ref. 203. The centerline velocity variation for heated and unheated cases with different nozzle sizes [and hence different *Fr*; see Eq. (178)] is given here as Fig. 178. The important effect of *Fr* is evident. The profiles of

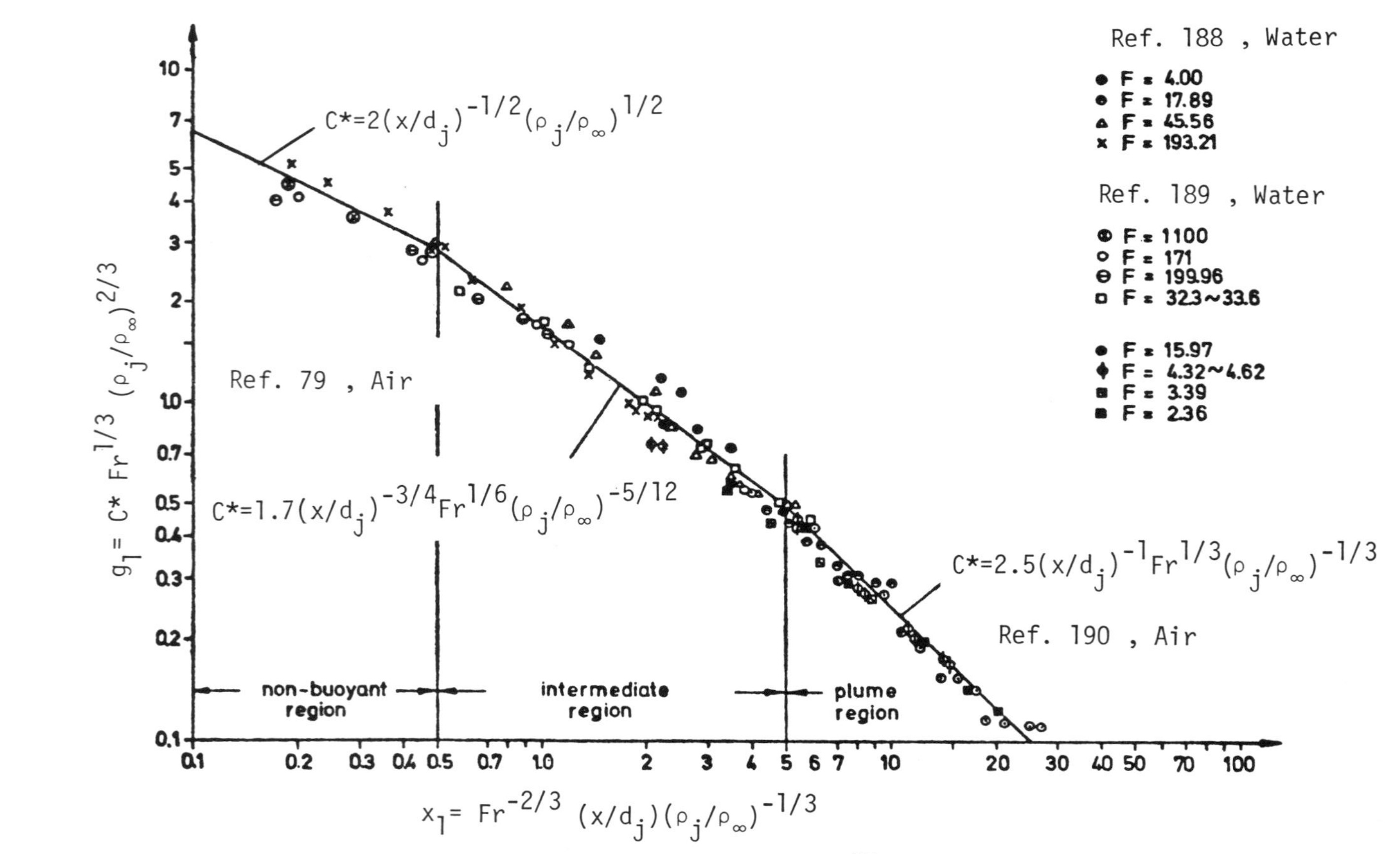

Fig. 173 Decay of centerline density in planar buoyant jets[187] (published with permission).

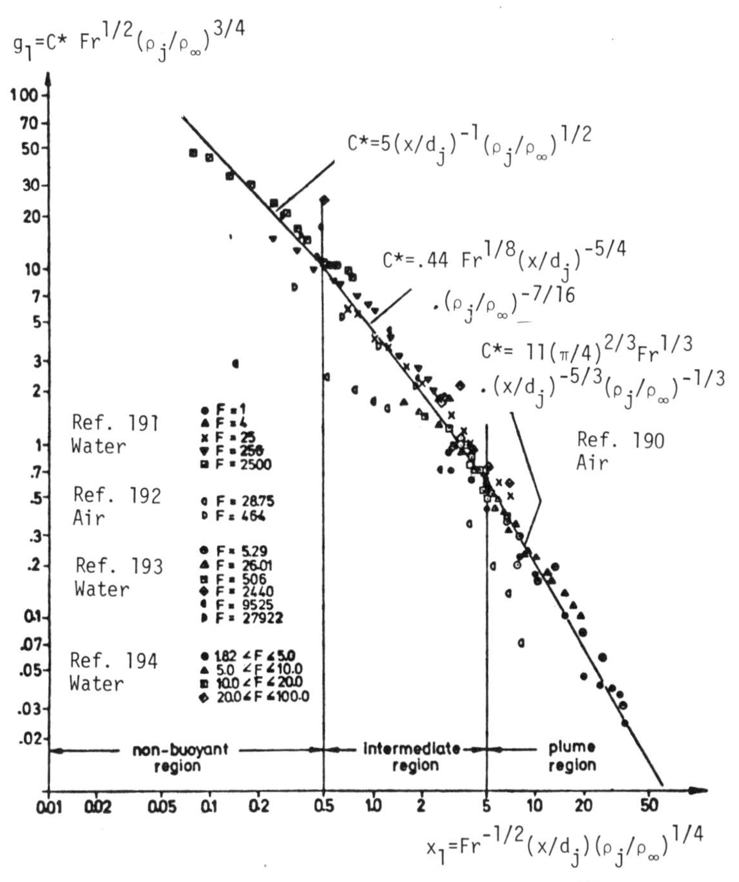

Fig. 174 Decay of centerline density in axisymmetric buoyant jets[187] (published with permission).

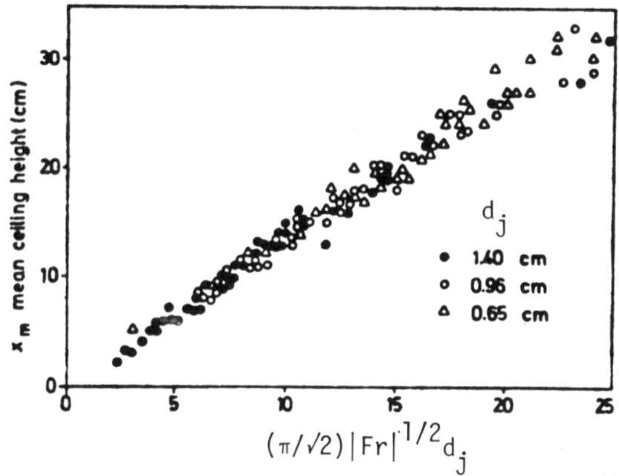

Fig. 175 Negative buoyancy axisymmetric jet in uniform surroundings[195] (published with permission).

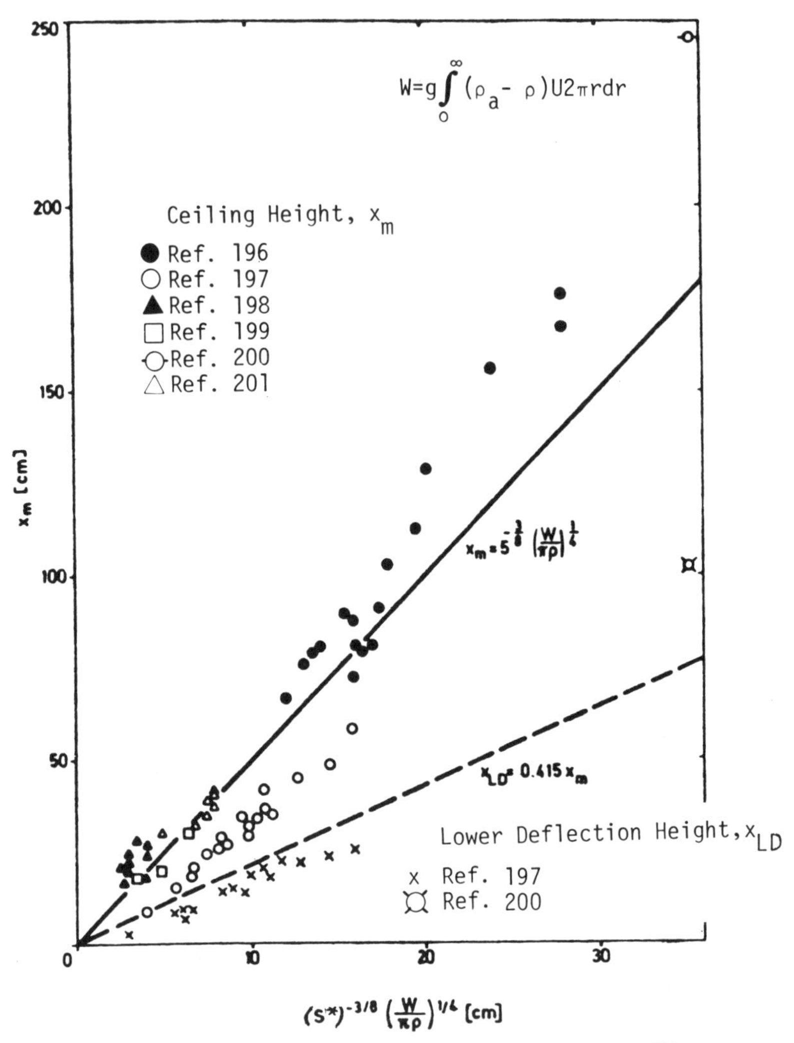

Fig. 176 Buoyant axisymmetric jets in a linearly stratified environment[187] (published with permission).

velocity and temperature in this case show that the influence of buoyancy forces is to make the temperature profile narrower than the velocity profile (see Fig. 179), which is opposite to the nonbuoyant jet case. The turbulence data shown in Fig. 180 indicate no appreciable effect of *Fr* in this range.

2. *Wakes in a Stratified Environment*

The behavior of a horizontal wake in a vertically stratified medium has been the subject of a number of experiments, and many of the results obtained were

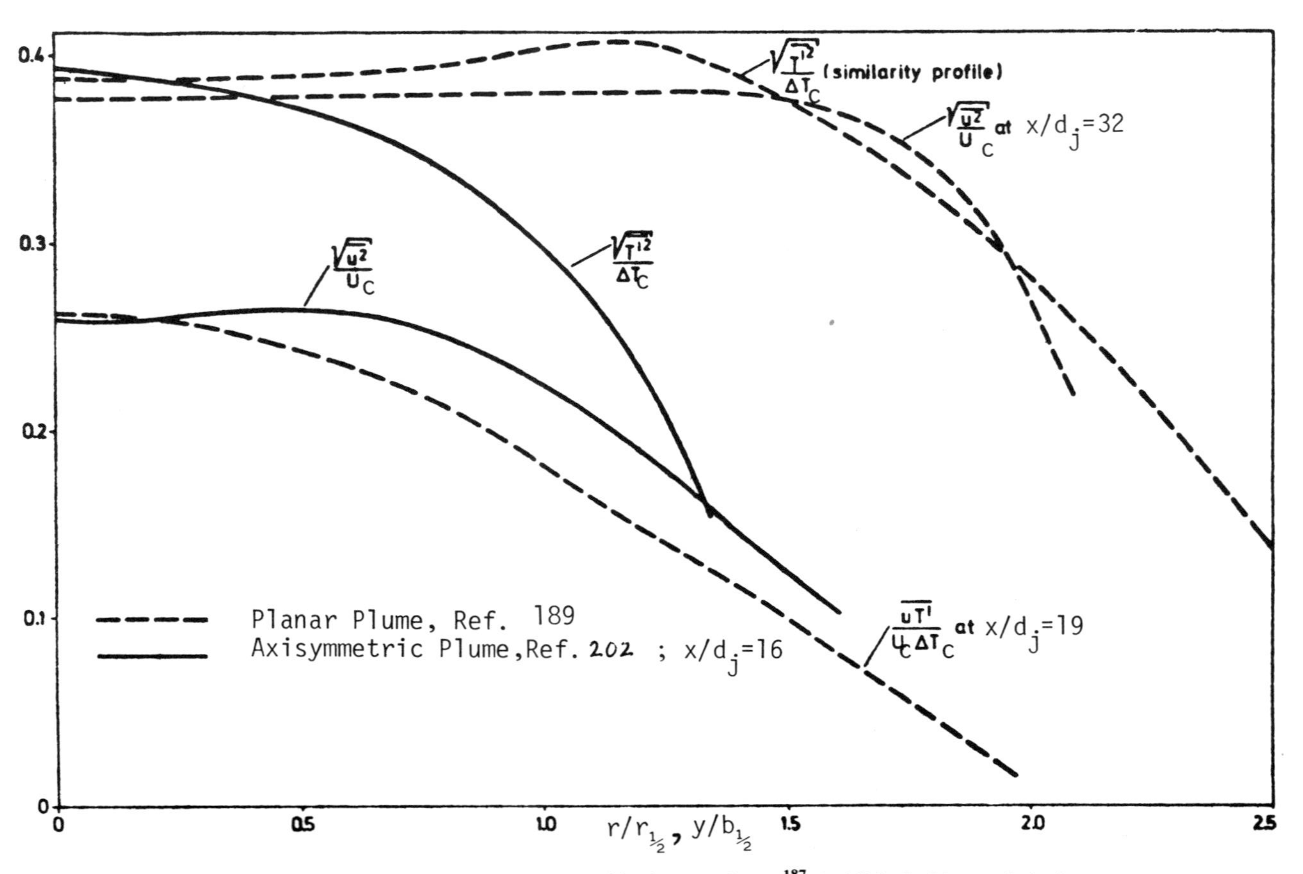

Fig. 177 Profiles of turbulence quantities in pure plumes[187] (published with permission).

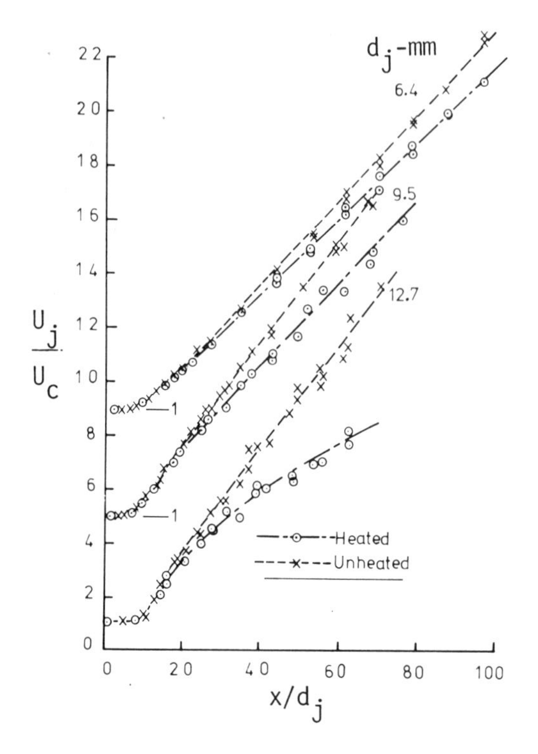

Fig. 178 Centerline velocity variation in buoyant axisymmetric jets[203] (published with permission).

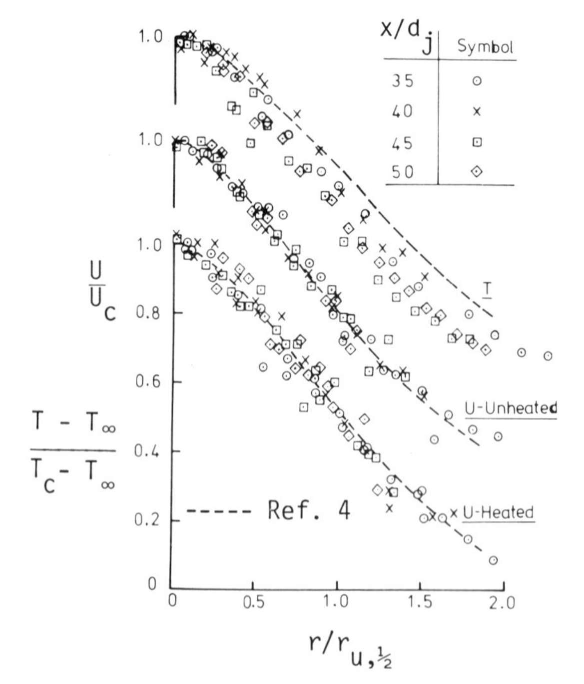

Fig. 179 Radial profiles of velocity in buoyant axisymmetric jets[203] (published with permission).

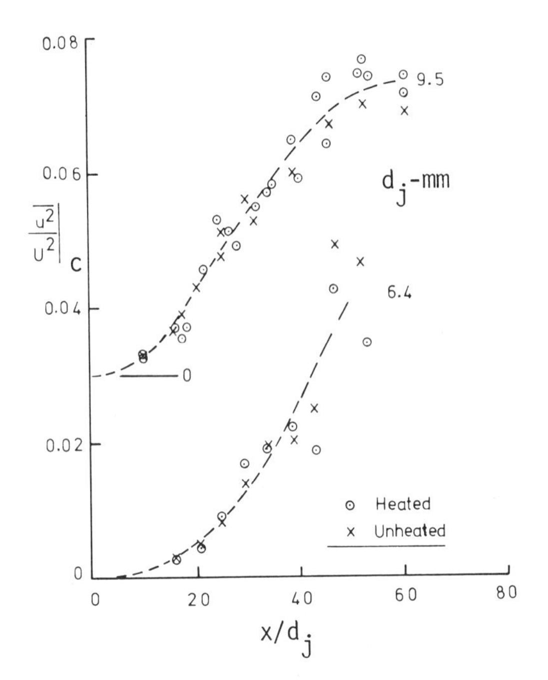

Fig. 180 Streamwise variation of centerline turbulence intensity in buoyant axisymmetric jets[203] (published with permission).

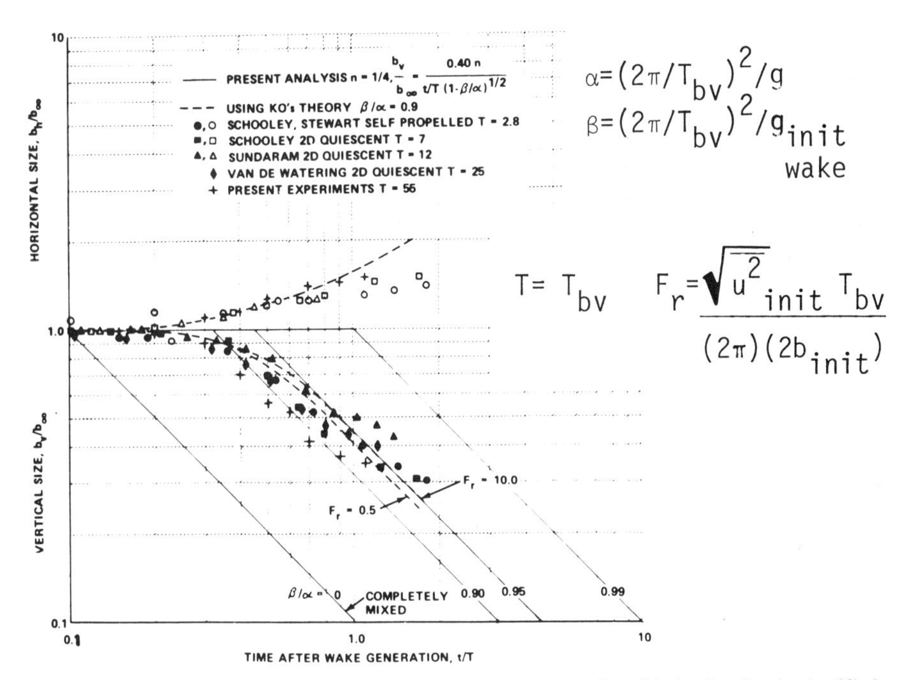

Fig. 181 Correlation of wake dimensions during collapse for various kinds of wakes in stratified fluids.[204]

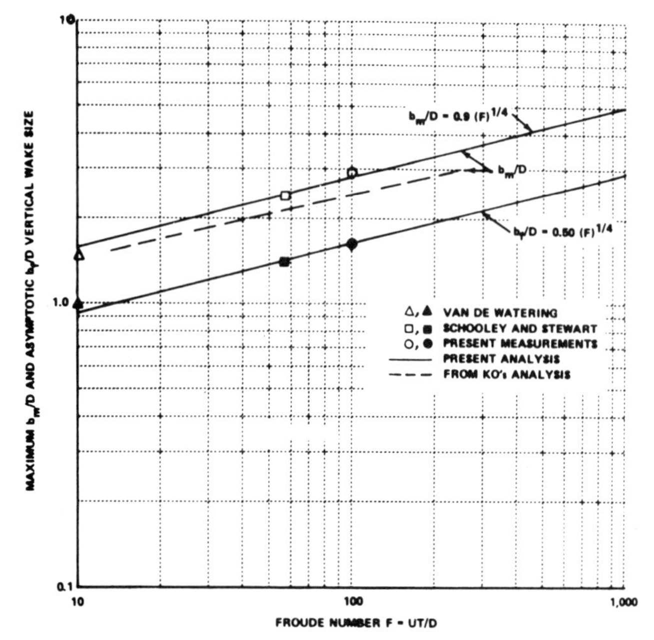

Fig. 182 Correlation of maximum and asymptotic vertical width for wakes in stratified fluids.[204]

surveyed and correlated in Ref. 204. The most obvious influence of stably stratified surroundings is a marked inhibition of the growth of the wake in the vertical direction with downstream distance. This phenomenon has been called "wake collapse," although the term implies somewhat more dramatic events than are actually observed for the wake behind real bodies. The width of the wake in the horizontal plane grows essentially as that in an unstratified medium. Results for these two wake widths are plotted on Fig. 181 in terms of the downstream distance parameter (t/T_{bv}), where T_{bv} is the Brunt-Vaisala period:

$$T_{bv} \equiv \frac{2\pi}{[-g/\rho(\partial\rho/\partial z)]^{1/2}} \tag{182}$$

and z is measured in the vertical direction. The "analysis" curves are discussed in Ref. 204. The maximum and asymptotic vertical extent of the wake is shown in Fig. 182.

C. Analysis

1. Mean-Flow Models

The integral analysis for transverse jet flows of Ref. 158 was extended to include the effects of buoyancy forces and stratification of the surroundings. For the case of a horizontal, buoyant jet injected across a horizontal mainstream, a three-dimensional jet trajectory is produced as the mixing zone rises (or falls) due to buoyancy and is bent over in the horizontal plane due to the

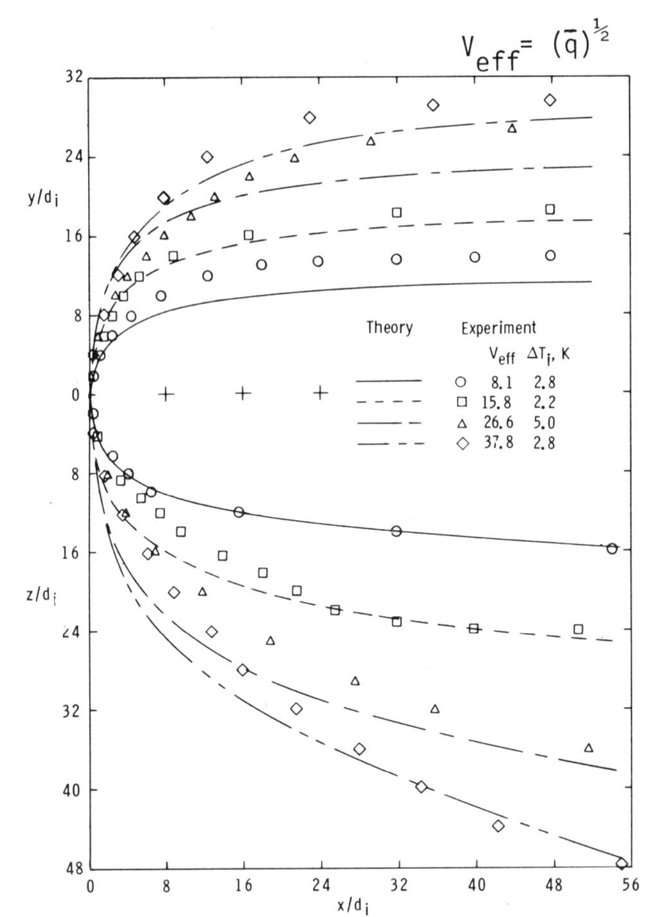

Fig. 183 Comparison of prediction and experiment for trajectory analysis of buoyant transverse jets[158]; the jet is injected across a horizontal stream, $\theta = 90$ deg, and up, $\gamma = 50$, from the horizontal plane.

crossflow. Comparison of prediction and experiment for some cases of that type is shown in Fig. 183. The jet was injected across a horizontal stream at an angle of 50 deg to the horizontal. Other integral analyses where the jet trajectory is restricted to only a vertical plane are given in Refs. 198, 200, and 209.

The numerical Navier-Stokes analysis of Ref. 186 was applied to the case of a buoyant, horizontal jet in a horizontal crossflow. Some results are shown in Fig. 184, where it can be seen that the influence of buoyancy in producing plume rise was somewhat underestimated. The prediction of cross-stream penetration is in better agreement with experiment.

A combined differential-integral type of analysis is presented in Ref. 210. Some comparisons of predictions with the data of Ref. 211 for the relatively simple case of a negatively buoyant, horizontal jet in a coaxial ambient flow are shown in Fig. 185.

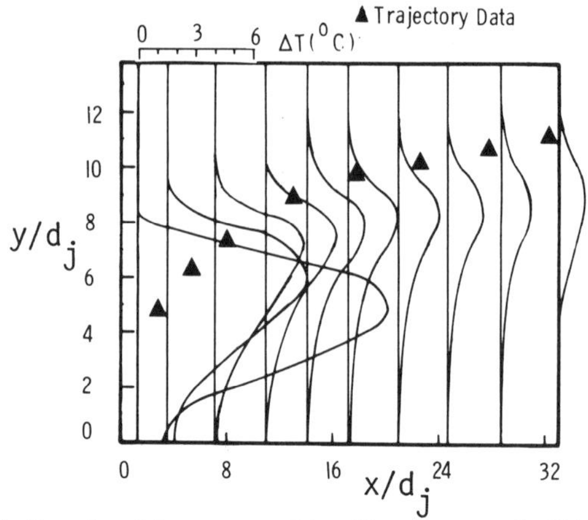

a) Projections on the Horizontal Plane

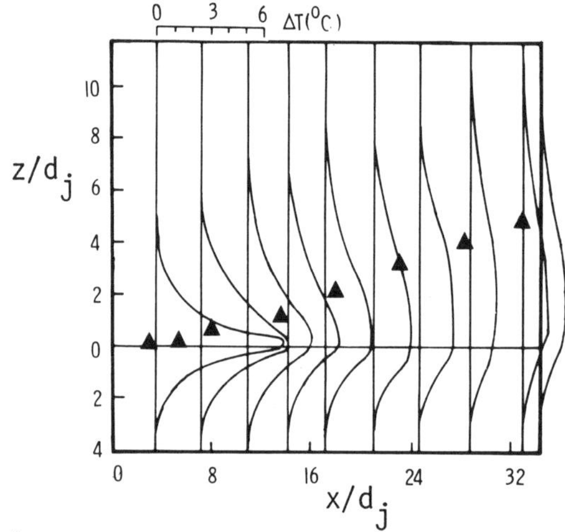

b) Projections on the Vertical Plane

Fig. 184 Comparison of predicted temperature profiles based on a mean-flow model within the Navier-Stokes equations and trajectory data for a heated horizontal axisymmetric jet transverse to a horizontal stream.[186]

2. *Two-Equation Models*

Beyond the mean-flow models, most workers have used the k-ϵ (or k-ϵ-$\overline{T'^2}$) level of model.[212-215] The analysis of Ref. 213 is essentially an extension of the $k\epsilon 2$ model of Ref. 81 to buoyant flows; however, the eddy viscosity concept is used only indirectly in favor of direct algebraic relationships between turbulent stresses and heat fluxes, k, and the mean velocity and temperature fields. These relations are developed following the original suggestions in Ref. 34. First, the transport equations for $\overline{u_i u_j}$ and $\overline{u_i T'}$ are written as

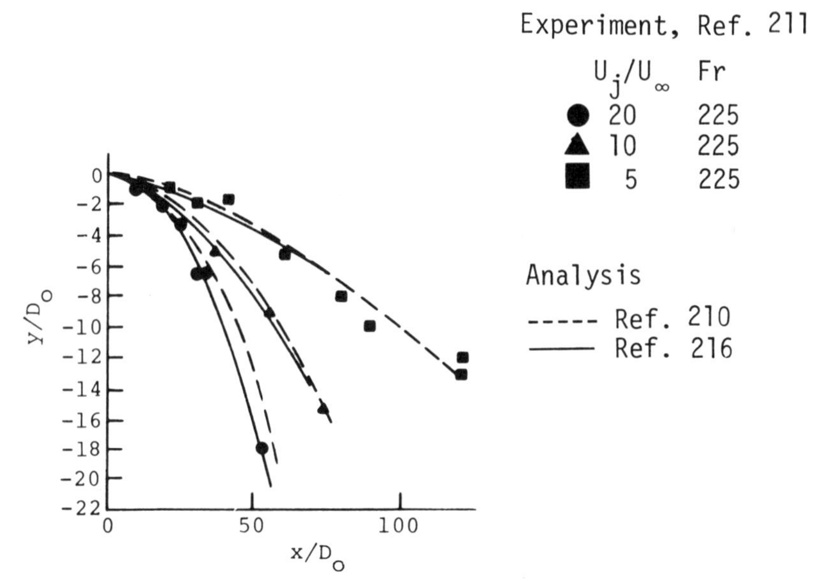

Fig. 185 **Comparison of prediction with a mean-flow model and the experiments of Ref. 211 for a heated horizontal axisymmetric jet in a coflowing stream[210] (published with permission).**

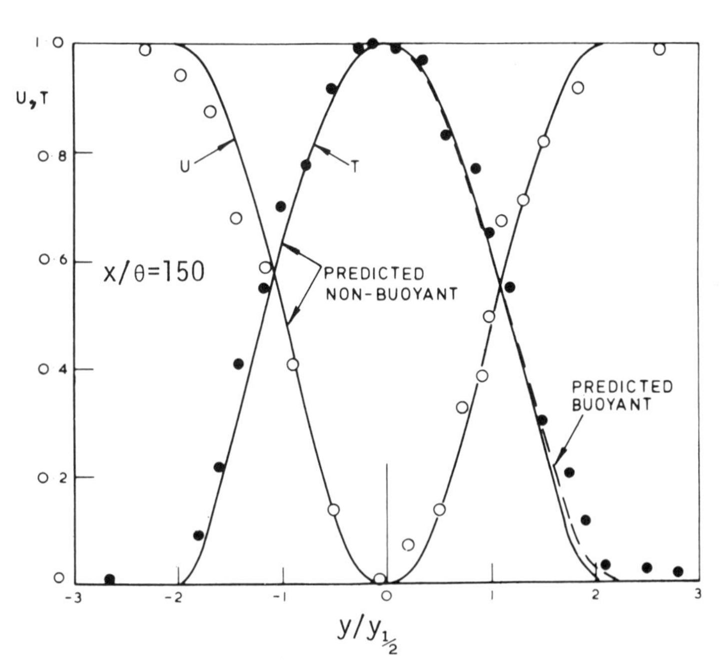

Fig. 186 **Comparison of predictions with a Reynolds stress model from Ref. 213 and the experiments of Refs. 217 and 218 for the wake behind a heated plate: mean flow (published with permission).**

$$\frac{\mathrm{D}}{\mathrm{D}t}(\overline{u_i u_j}) = \mathfrak{D}(\overline{u_i u_j}) + \mathcal{P}_{ij} - \frac{2}{3}\delta_{ij}\epsilon - C_1 \frac{\epsilon}{k}\left(\overline{u_i u_j} - \frac{2}{3}\delta_{ij}k\right) - C_2\left(\mathcal{P}_{ij} - \frac{2}{3}\delta_{ij}\mathcal{P}\right) \quad (183)$$

$$\frac{\mathrm{D}}{\mathrm{D}t}(\overline{u_i T'}) = \mathfrak{D}(\overline{u_i T'}) + \overline{u_i u_k}\frac{\partial T}{\partial x_k} - C_{1T}\frac{\epsilon}{k}(\overline{u_i T'}) + \mathcal{P}_{iT}(1 - C_{2T}) \quad (184)$$

Here δ_{ij} is the Kronecker delta, $\mathfrak{D}(\ \ )$ denotes the rate of diffusive transport of ( ), and $\mathcal{P}$, $\mathcal{P}_{ij}$, and $\mathcal{P}_{iT}$ are the production rate of k, $\overline{u_i u_j}$, and $\overline{u_i T'}$. Next, following Ref. 34, Eq. (183) is approximated as

$$\left[\frac{\mathrm{D}}{\mathrm{D}t} - \mathfrak{D}\right](\overline{u_i u_j}) = \frac{\overline{u_i u_j}}{k}\left[\frac{\mathrm{D}}{\mathrm{D}t} - \mathfrak{D}\right](k) \quad (185)$$

With the equation for k, this becomes

$$\left[\frac{\mathrm{D}}{\mathrm{D}t} - \mathfrak{D}\right](\overline{u_i u_j}) = \frac{\overline{u_i u_j}}{k}(\mathcal{P} - \epsilon) \quad (186)$$

Combining Eqs. (186) and (183) gives

$$\frac{\overline{u_i u_j} - (2/3)\delta_{ij}k}{k} = \left(\frac{1 - C_2}{C_1 - 1 + \mathcal{P}/\epsilon}\right)\left(\frac{\mathcal{P}_{ij} - (2/3)\delta_{ij}\mathcal{P}}{\epsilon}\right) \quad (187)$$

A similar development led to

$$-\overline{u_i T'} = \frac{(k/\epsilon)\overline{u_i u_k}(\partial T/\partial x_k) - (1 - C_{2T})(k/\epsilon)\mathcal{P}_{iT}}{C_{1T} + [(\mathcal{P}/\epsilon - 1)/2]} \quad (188)$$

Finally, it was assumed that

$$\overline{T'^2} = -C_T' \frac{k}{\epsilon}\overline{u_k T'}\frac{\partial T}{\partial x_k} \quad (189)$$

and $C_T' = 1.6$, which means that a separate transport equation for this quantity was not needed.

A restricted form of this model for thin, horizontal flows was applied to some experimental cases in Ref. 213. The $\overline{\mathcal{P}}/\epsilon$ function used differs somewhat from that in Ref. 81. Mean velocity and temperature profiles for the wake flow of Refs. 217 and 218 are shown in Fig. 186. The influence of buoyancy under these conditions is seen to be slight. That is not the case, however, for the turbulent shear stress and heat flux, as seen in Fig. 187. The analysis clearly predicts the data quite well.

A related, but more simplified, turbulence model was applied to the case of a round, buoyant, vertical jet into neutral or stratified surroundings in Ref. 214. In that case, however, an additional equation was solved for the variation of $\overline{T'^2}$. A comparison of a prediction with data for an experimental case from Ref. 200 is presented here as Fig. 188. Good agreement is apparent.

The k-ϵ model of Ref. 216 with some further approximations was applied to the problem of a wake in a stratified medium in Ref. 219. Comparison with the self-propelled data of Refs. 220-223 is shown in Fig. 189a in terms of the

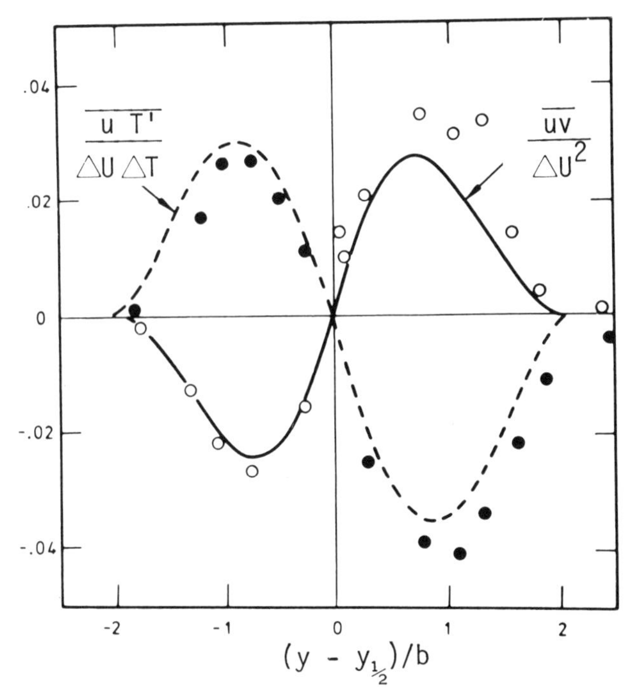

Fig. 187 Comparison of predictions with a Reynolds stress model from Ref. 213 and the experiments of Refs. 217 and 218 for the wake behind a heated plate: turbulent transport (published with permission).

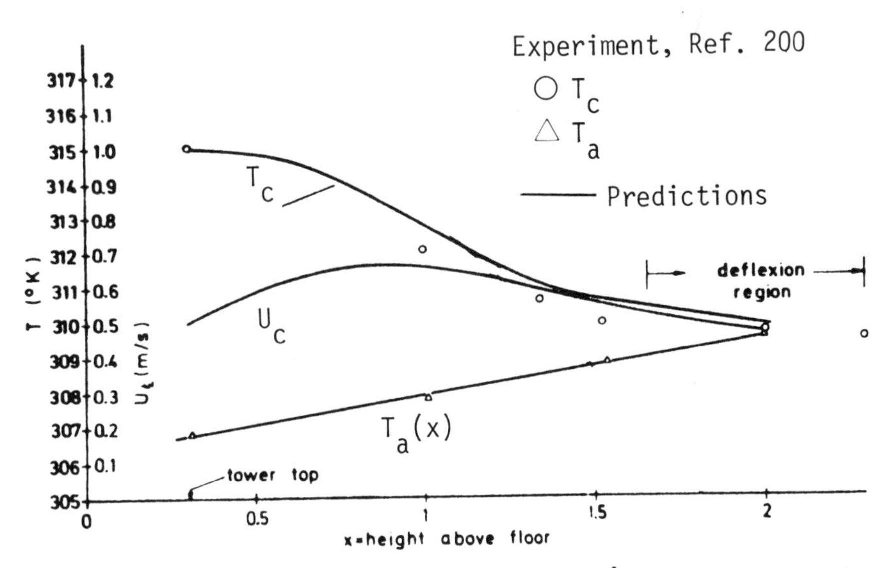

Fig. 188 Comparison of prediction with a three-equation ($k\epsilon T'^2$) model and experiment for a vertical plume of Ref. 200[214] (published with permission).

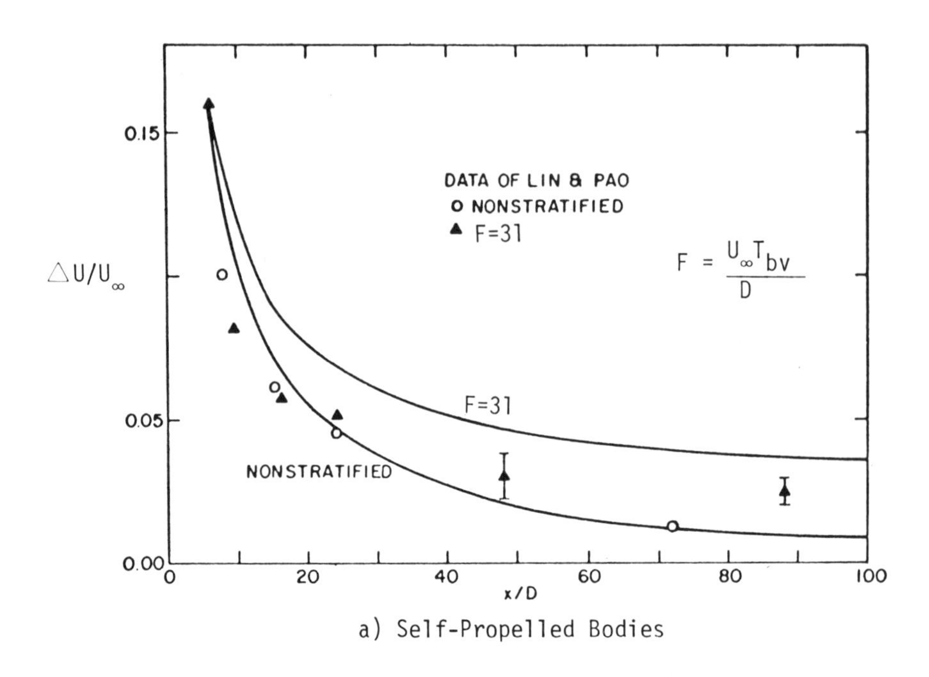

a) Self-Propelled Bodies

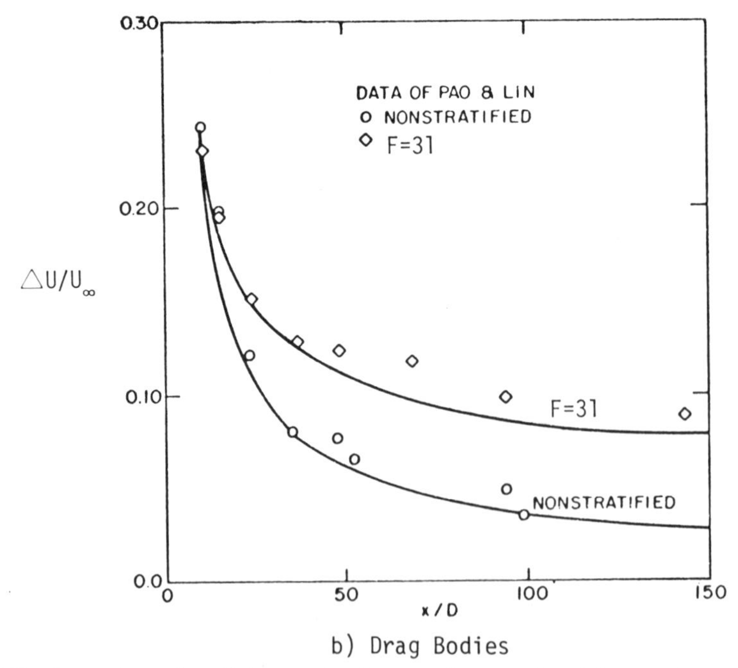

b) Drag Bodies

Fig. 189 Comparison of predictions with a two-equation model and the experiments of Refs. 220-223 for wakes in stratified surroundings.[219]

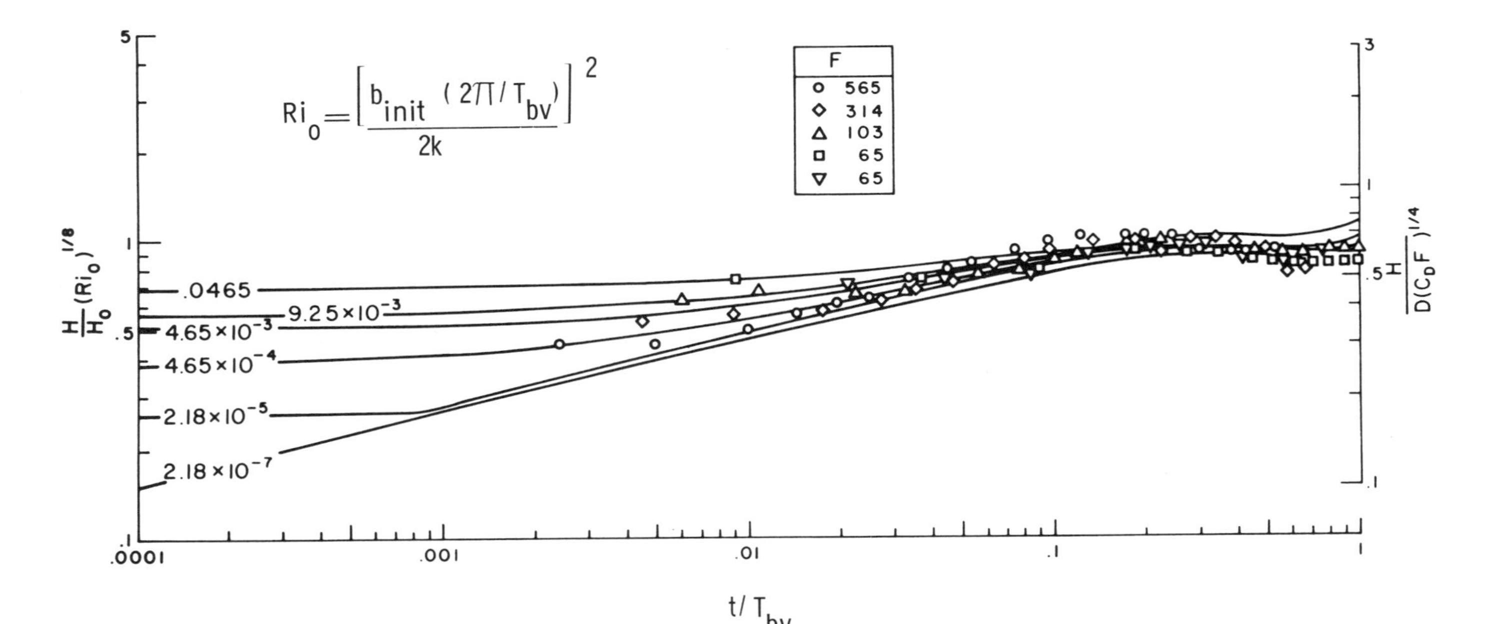

Fig. 190 Comparison of predictions with a Reynolds stress model from Ref. 224 and the experiments of Ref. 220 for the wake behind bodies in stratified surroundings. The scale for the predictions is on the left, and the data scale is on the right, with the scales matched at the first local maximum of H.

centerline velocity variation. The effects of stratification are overpredicted. Agreement with experiment for the drag body case is much better, as shown in Fig. 189b.

3. Reynolds Stress Models

The Reynolds stress model of Donaldson and co-workers has been applied to stratified flows of various types, including wakes, in Ref. 224. A comparison of the growth of the wake in the vertical direction from the experiments of Ref. 220 and as predicted by the analysis is shown in Fig. 190. Although this comparison is only for a gross descriptor of the flowfield, the agreement with experiment is good.

X. VISCOUS-INVISCID INTERACTIONS

A. Background

We have already seen that, in most cases, whether the flow is subsonic or supersonic does not play an important role in mixing analysis. The one major exception to this point has been the case of transverse injection into a supersonic flow where the formation of an interaction shock was important. A second major category, related to the first, encompasses the cases with a supersonic main flow and where the mixing process itself produces significant flow deflections. Since flow deflections in supersonic flow are explicitly related to substantial static pressure changes, these effects must be adequately treated in analysis. A common practical example for this situation is injection at a low rate through the base of a body moving at supersonic speeds. Of course, the same type of flowfield results from jet injection at a low rate through a simple nozzle when the external flow is supersonic. The importance of the phrase "at a low rate" emerges, since the initially low-speed flow in the mixing zone is accelerated by the external stream via turbulent shear until it passes into the supersonic regime through a "viscous throat." This process is controlled by the viscous-inviscid interaction of flow deflection and pressure change on the boundary of the mixing zone.

Similar general flow processes are obtained with no injection, and those cases are commonly called "base flows." A separate volume in this series[225] was recently published on some aspects of that general subject, and only cases with base injection and newer work will be considered here.

B. Analysis

The pioneering work on this general subject where the viscous-inviscid interaction was treated directly was the integral analysis published in Ref. 226. The works of Refs. 227 and 228 made further extensions and applications of the same basic analytical model. The main contributions of these studies were in the identification and treatment of the saddle-point singularity at the viscous throat. Turbulent transport phenomena entered through an entrainment law, which was taken in Ref. 226 as

$$\frac{\mathrm{d}\dot{m}}{\mathrm{d}x} = 0.03\rho_\infty U_\infty \tag{190}$$

Entrainment occurs through a difference between the local angle of a streamline and the orientation of the outer edge of the mixing zone. For a planar case (the only cases treated in Refs. 226-228), this may be written as

$$\frac{\mathrm{d}\dot{m}}{\mathrm{d}x} = \rho_\infty U_\infty \left(\frac{\mathrm{d}b}{\mathrm{d}x} - \theta \right) \tag{191}$$

Combining with Eq. (190) gives

$$\theta = \frac{db}{dx} - 0.03 \tag{192}$$

If, for example, the outer supersonic flow is taken as adequately described by classical linear supersonic theory, the pressure change accompanying this flow deflection can be expressed by

$$\frac{dP}{P} = \frac{\gamma M^2 \theta}{\sqrt{M^2 - 1}} \tag{193}$$

This static pressure variation is imposed on the whole mixing zone and obviously cannot be neglected.

In their present form, the analyses of Refs. 226-228 are restricted to isoenergetic flows and planar geometry, and, as stated by the original authors, the removal of either of these restrictions is by no means a trivial exercise. They also do not consider injection. Reference 227 reports an attempt to treat base injection by the Alber-Lees approach, but poor results were obtained. Detailed numerical treatments based upon the Navier-Stokes equations have enjoyed only limited success for much simpler problems of the same general type.[225] Moreover, what limited success has been obtained was purchased at a very dear price in terms of programming and computer execution time.

On the other hand, base injection in planar cases using a simplified treatment of the Crocco-Lees type has been treated in Refs. 229 and 230, and the results are in good agreement with experiment. For the injection problem, it

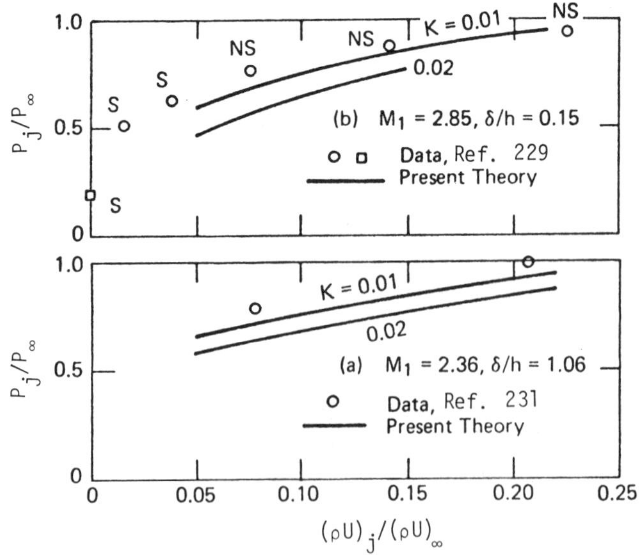

Fig. 191 Comparison of prediction of an approximate theory[230] and experiments of Refs. 229 and 231 for planar base flow with injection.

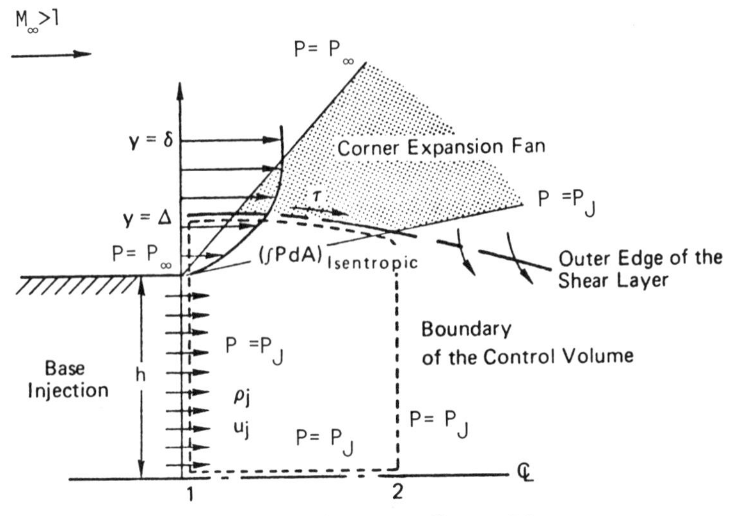

Fig. 192 Schematic of the corner-flow model.

appeared necessary to extend Eq. (190) to account for the small entrainment expected for the situation in which $\rho_\infty U_\infty = (\rho U)_{av}$; i.e., the mass flux in the freestream and the inner viscous zone are the same. Most successful eddy viscosity models predict a zero Reynolds stress for the same condition. Thus, the following was adopted:

$$\frac{d\dot{m}}{dx} = K\rho_\infty U_\infty \left| 1 - \frac{(\rho U)_{av}}{\rho_\infty U_\infty} \right| \tag{194}$$

A perhaps surprising result of the analysis is that one cannot arbitrarily set the injection pressure for a given mass-flow rate of injectant. That pressure plays the same role here as the base pressure does in the base-flow problem. The injection pressure is thus a result of the viscous-inviscid interaction, and the prediction of its value is an important test of any analysis. Some results from Ref. 230 are compared with experiment in Fig. 191. Here, the experiments of Refs. 229 and 231, which are slot injection configurations, have been used. However, the only difference between slot injection and base flow is the generally negligible effect of surface shear on the solid surface that replaces the body axis. The use of $K = 0.01$ yields the best agreement with the data for most values of $(\rho U)_j/\rho_\infty U_\infty$. At very low or no base flow, $K = 0.02$ gives better agreement. This may, in part, be due to whether or not substantial recirculation zones exist. The "S" or "NS" on Fig. 191 indicates whether or not separation zones were observed. The value $K = 0.01$ was chosen for general application for problems with injection.

The analytical treatment of the flow at the jet lip or the "corner-flow" region is important, since it governs the initial conditions of the shear layer. Older experiments and numerical studies had indicated that it might be suitable simply to assume that only the subsonic portion of the body boundary

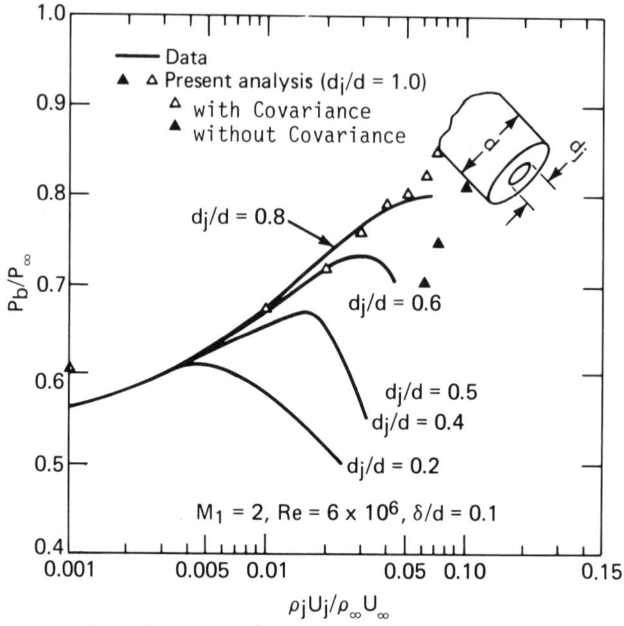

Fig. 193 Comparison of prediction of an approximate analysis for base pressure on an axisymmetric body with the experiments of Ref. 233.

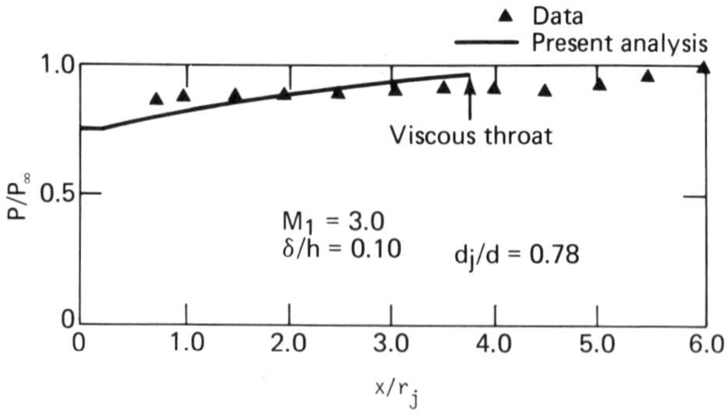

Fig. 194 Comparison of prediction of an approximate analysis for pressure distribution behind an axisymmetric body with injection with the experiment of Ref. 234 for $\rho_j U_j/\rho_\infty U_\infty = 0.037$.

layer participated in the shear-layer formation. An attempt to improve upon that rather crude treatment of this essential part of the total flow was presented in Ref. 228; however, their approach has been found to include an error in the formulation of the $\int P dA$ term. Reference 230 adopted a new model of the corner-flow problem, shown schematically in Fig. 192. It was presumed that some portion Δ of the original boundary-layer thickness δ participates in the formation of the shear layer. The Mach number at Δ is denoted by M_D, and no assumption as to its value is introduced. The analysis considers a control volume as indicated by the dotted line and applies conservation of mass, momentum, and energy. Any value of base injection, including zero, is admissible. A double iterative process is employed to determine a solution.

After a problem is specified by given values of M_∞, δ/h, $\rho_j U_j$, and M_j, a series of trial values for P_j/P_∞ is selected. For each trial value of P_j/P_∞, a series of trial values for M_D is selected. For each resulting case, a solution to the equations of motion is sought. This results in a series of solutions to a given problem specified by pairs of values of $(\rho U)_{av}/\rho_\infty U_\infty$ and P_j/P_∞. The correct solution is selected by matching with the rest of the flowfield solution, as will be illustrated below.

Several important assumptions are incorporated into the solution of the equations of motion which produces the pairs of $(\rho U)_{av}/\rho_\infty U_\infty$ and P_j/P_∞. First, the PdA term in the momentum equation is taken as that for an isentropic turn through the same pressure ratio P_j/P_∞. Also the geometry of the system is obtained via the isentropic assumption. Second, the shear along the outer boundary is determined by evaluating the slope at $y=\Delta$ and employing a compressible, outer-region eddy viscosity model. Third, the boundary-layer profile is taken as a simple, one-seventh power law.

Finally, this type of approximate analysis has been extended to axisymmetric cases by the use of a heuristic pressure-flow deflection relationship in Ref. 232. Some comparison with data is shown in Figs. 193 and 194, where the reasonable agreement with data can be seen.

XI. CLOSURE

For all of the experimental studies, it is often helpful for the reader to understand fully the apparatus and the techniques employed in order to assess the accuracy of the data. Also, some information is clearly helpful to design new experiments to verify older data, extend the range of available data, or test new turbulence models. It is beyond the scope of this volume to provide an in-depth discussion of these topics, but a brief bibliography of some major sources was deemed worthwhile.

References 235-237 are good general references on wind tunnels and related apparatus. For the mean-flow variables, the most common measurement is (U, V, W), and this is usually accomplished with measurements of dynamic, static, and/or total pressure and flow angularity. References 235-239 provide thorough treatments. The measurement of (T) is almost always made with thermocouples, and Refs. 239 and 240 contain the necessary details. Mean concentration (C) data are most commonly obtained by sampling and either subsequent or on-line gas analysis. The techniques of Ref. 49 are representative of modern approaches. In two-phase cases, the design of the probes becomes especially important, and the techniques of Refs. 141 and 145 are typical.

Data on the fluctuating quantities (u, v, w, T', c) and their correlations (e.g., $\overline{uv}, \overline{uT'}$) are still made in most cases with various configurations of the hot-wire anemomenter. References 242 and 243 are readable general references on the subject. Reference 47 is a recent example of modern techniques for (T') and $(\overline{uT'})$. The measurement of (c) with hot wires is explored fully in Ref. 59. Optical methods have long been popular for determining the gross behavior of high-speed flows. Mean-flow data can be obtained with some of the methods, and it is also possible to obtain some types of turbulence information through relatively simple optical techniques. A discussion is given in Ref. 240. The laser Doppler velocimeter (LDV) has been receiving ever wider application, especially for low-speed and liquid flows. References 244-246 contain background information and the current view of the problems, as well as the promise of this general method for both mean and turbulence velocities. The area of greatest difficulty is still that of measuring fluctuating quantities containing the concentration. The Raman scattering methods now seem to be near the point of routine application, as shown by the results of Ref. 60.

Looking back over the experimental and analytical results that have been presented, it becomes apparent that the state of knowledge in injection and mixing flows has not yet developed to the point where clear choices for experimental methods or turbulence models can be made for a new flow problem. For the development and/or verification of turbulence models, in particular, there are large and unfortunate holes in the coverage of available data. Too often, the data from separate experiments for presumably identical (or nearly so) flow problems do not agree. We must gain a deeper understanding of the factors that produce these disagreements. The problem is most severe for two-phase flows, but troubling disagreements exist even in the simplest flow situations. Lastly, the comparison of predictions based upon various turbulence models from different workers with data for the same flow problem from different sources is very equivocal. The difficulty of relative comparison is sometimes aggravated by the use of different numerical approaches, different approximations to the equations of motion, and slightly different values of proportionality constants and/or such parameters as the turbulent Prandtl or Schmidt numbers even for ostensibly identical turbulence models. A final plea here is for the publication of understandable statements of computer time with each computation. This is most certainly not a matter of secondary importance to the potential user.

REFERENCES

[1]Schlichting, H., *Boundary Layer Theory,* McGraw-Hill, New York, various editions since 1942.

[2]Pai, S. I., *Fluid Dynamics of Jets,* D. Van Nostrand, New York, 1954.

[3]Birkhoff, G. and Zarantello, E.H., *Jets, Wakes and Cavities,* Vol. 2, *Applied Mathematics and Mechanics*, Academic Press, New York, 1957.

[4]Hinze, J. L., *Turbulence*, McGraw-Hill, New York, 1959.

[5]Abramovich, G. N., *The Theory of Turbulent Jets,* MIT Press, Cambridge, Mass., 1960 (English edition).

[6]Ferri, A., Moretti, G., and Slutsky, S., "Mixing Processes in Supersonic Combustion," *Journal of the Society of Industrial and Applied Mathematics*, Vol. 13, March 1965.

[7]Brodkey, R. S., "Fluid Motion and Mixing," *Mixing, Theory and Practice,* edited by V. S. Uhl and J. B. Gray, Academic Press, New York, 1966.

[8]Ferri, A., "A Critical Review of Heterogeneous Mixing Problems," Aeronautical Research Lab., ARL-67-0187, Sept. 1967.

[9]Newman, B. G., "Turbulent Jets and Wakes in a Pressure Gradient," *Fluid Mechanics of Internal Flow,* edited by G. Sovran, Elsevier Publishing Co., Amsterdam, 1967.

[10]*Free Turbulent Shear Flows,* NASA SP-321, 1973.

[11]Bradbury, L. J. S., "A Review of Theoretical and Experimental Work on the Development of Turbulent Jets," *Turbulent Jet Flows,* Von Karman Inst., Rhode-Saint-Genese, Belgium, VKI-LS-36, 1971.

[12]Harsha, P. T., "Free Turbulent Mixing: A Critical Evaluation of Theory and Experiment," *Turbulent Shear Flows,* AGARD CP-93, 1971.

[13]Bradshaw, P., "The Understanding and Prediction of Turbulent Flow," *Aeronautical Journal,* July 1972.

[14]Tennekes, H. and Lumley, J. L., *A First Course in Turbulence,* MIT Press, Cambridge, Mass., 1972.

[15]Launder, B. and Spalding, D. B., *Lectures in Mathematical Models of Turbulence*, Academic Press, London, 1972.

[16]Murthy, S. N. B. (ed.), *Turbulent Mixing in Nonreactive and Reactive Flows*, Plenum Press, New York, 1974.

[17]Murthy, S. N. B., "Turbulent Mixing in Nonreactive and Reactive Flows—A Review," *Turbulent Mixing in Nonreactive and Reactive Flows,* edited by S. N. B. Murthy, Plenum Press, New York, 1974.

[18]Abramovich, G. N., *Turbulent Mixing of Gaseous Jets*, Moscow, USSR, 1974 (in Russian).

[19]Abramovich, G. N., Krasheninikov, S. U., and Sekundov, A. N., *Turbulent Flows with Body Forces and Non-Similarity,* Moscow, USSR, 1975 (in Russian); see also an article with the same title in English in *Fluid Mechanics—Soviet Research*, Vol. 5, Sept. 1976.

[20]Bradshaw, P., "Review—Complex Turbulent Flows," *Journal of Fluids Engineering*, June 1975.

[21]Rodi, W., "A Review of Experimental Data of Uniform Density Free Turbulent Boundary Layers," *Studies in Convection,* Vol. 1, edited by B. E. Launder, Academic Press, London, 1975.

[22]Rajaratham, M., *Turbulent Jets*, Vol. 5, *Developments in Water Science,* Elsevier Science Publishing Co., Amsterdam, 1976.

[23]Bradshaw, P. (ed.), *Turbulence*, Vol. 12, *Topics in Applied Physics,* Springer-Verlag, Berlin, 1978.

[24]Townsend, A. A., *The Structure of Turbulent Shear Flow,* 2nd ed., Cambridge University Press, England, 1976.

[25]Bradshaw, P., Ferriss, D. H., and Atwell, N. P., "Calculation of Boundary Layer Development Using the Turbulence Energy Equation," *Journal of Fluid Mechanics,* Vol. 28, Pt. 3, 1967, pp. 539-616.

[26]Lumley, J. L., "The Applicability of Turbulence Research to the Solution of Internal Flow Problems," *Fluid Mechanics of Internal Flow,* edited by G. Sovran, Elsevier Publishing Co., Amsterdam, 1967.

[27]Antonia, R. A. and Bilger, R. W., "An Experimental Investigation of an Axi-symmetric Jet in a Co-flowing Air Stream," *Journal of Fluid Mechanics,* Vol. 61, Pt. 4, 1973, pp. 805-822.

[28]Weinstein, A. S., Osterle, J. F., and Forstall, W., "Momentum Diffusion from a Slot Jet into a Moving Secondary," *Journal of Applied Mechanics*, Sept. 1956, pp. 437-443.

[29]Bradbury, L. J. S., "The Structure of a Self-Preserving Turbulent Plane Jet," *Journal of Fluid Mechanics*, Vol. 23, Pt. 1, 1965, pp. 31-64.

[30]Everitt, K. W. and Robins, A. G., "The Development and Structure of Turbulent Plane Jets," *Journal of Fluid Mechanics*, Vol. 88, Pt. 3, 1978, pp. 563-583.

[31]Hokenson, G. J. and Schetz, J. A., "Free Turbulent Mixing in Axial Pressure Gradients," *Journal of Applied Mechanics*, Vol. 40, Feb. 1973, pp. 375-380.

[32]Schetz, J. A., "Some Studies of the Turbulent Wake Problem," *Astronautica Acta,* Vol. 16, 1971, pp. 107-117.

[33]Gibson, M. M., "Spectra of Turbulence in a Round Jet," *Journal of Fluid Mechanics,* Vol. 15, Pt. 2, 1963, pp. 161-173.

[34]Rodi, W., "The Prediction of Free Turbulent Boundary Layers by Use of a Two-Equation Model of Turbulence," Ph.D. Thesis, Univ. of London, 1972.

[35]Wygnanski, I. and Fiedler, H. E., "Some Measurements in the Self-Preserving Jet," *Journal of Fluid Mechanics,* Vol. 38, 1969, pp. 577-612.

[36]Carmody, T., "Establishment of a Wake Behind a Disk," *Journal of Basic Engineering,* Vol. 87, 1964, pp. 869-882.

[37]Chevray, R., "The Turbulent Wake of a Body of Revolution," *Journal of Basic Engineering,* Vol. 90, 1968, pp. 275-284.

[38]Lawn, C. J., "The Determination of the Rate of Dissipation in Turbulent Pipe Flow," *Journal of Fluid Mechanics*, Vol. 48, 1971, pp. 477-505.

[39]Corrsin, S., "An Investigation of the Flow in an Axially Symmetric Heated Jet," NACA Rept. W-94, 1943.

[40]Corrsin, S. and Uberoi, M., "Further Experiments on the Flow and Heat Transfer in a Heated Turbulent Air Jet," NACA TN 1895, 1949.

[41]Landis, F. and Shapiro, A. H., "The Turbulent Mixing of Co-Axial Gas Jets," Heat Transfer and Fluid Mechanics Inst., Preprints and Papers, Stanford University Press, Stanford, Calif., 1951.

[42]Forstall, W., Jr. and Shapiro, A. H., "Momentum and Mass Transfer in Coaxial Gas Jets," *Journal of Applied Mechanics*, Vol. 72, 1950, pp. 339-408.

[43]Pabst, O., "Die Ausbreitung Heisser Grasstrahlen in Bewegter Luft, I Teil-Versuche in Kerngebeit," *Deutsche Luftfahrtforschung*, Aug. 1944.

[44]Reynolds, A. J., "The Variation of Turbulent Prandtl and Schmidt Numbers in Wakes and Jets," *International Journal of Heat and Mass Transfer*, Vol. 19, 1976, pp. 757-764.

[45]Sakipov, Z. B. and Temirbaev, D. J., "On the Ratio of the Coefficients of Turbulent Exchange of Mass and Heat in a Free Turbulent Jet," *Tepli i Massoperenos,* Vol. 2, 1965, pp. 407-413.

[46]Ginevskii, A. S., "Turbulent Nonisothermal Jets of a Compressible Gas of Variable Composition," *Promyshlennaya Aerodinamika*, No. 27, 1966, pp. 31-54.

[47]Chevray, R. and Tutu, N. K., "Intermittency and Preferential Transport of Heat in a Round Jet," *Journal of Fluid Mechanics*, Vol. 88, Pt. 1, 1978, pp. 133-160.

[48]Zakkay, V., Krause, E., and Woo, S. D. L., "Turbulent Transport Processes for Axisymmetric Heterogeneous Mixing," *AIAA Journal,* Vol. 2, Nov. 1964, pp. 1939-1947.

[49]Chriss, D. E., "Experimental Study of Turbulent Mixing of Subsonic Axisymmetric Gas Streams," Arnold Engineering Development Center, AEDC-TR-68-133, Aug. 1968.

[50]Abramovich, G. N., Yakovlevsky, O. V., Smirnova, I. P., Sekundov, A. N., and Krasheninnikov, S. Yu., "An Investigation of the Turbulent Jets of Different Gases in a General Steam," *Astronautica Acta,* Vol. 14, 1969, pp. 229-240.

[51]Schetz, J. A., "Analysis of the Mixing and Combustion of Gaseous and Particle Laden Jets in an Airstream," AIAA Paper 69-33, 1969.

[52]Keagy, W. R. and Weller, A. E., "A Study of Freely Expanding Inhomogeneous Jets," Heat Transfer and Fluid Mechanics Inst., 1949, pp. 89-98.

[53] Alpinieri. L. J., "Turbulent Mixing of Coaxial Jets," *AIAA Journal*, Vol. 2, Sept. 1964, pp. 1560-1567.
[54] Zakkay, V. and Krause, E., "Mixing Problems with Chemical Reaction," *Supersonic Flow, Chemical Processes and Radiative Transfer*, Pergamon Press, London, 1964.
[55] Boradachev, V. Y., reported in *The Theory of Turbulent Jets*, G. N. Abramovich, MIT Press, Cambridge, Mass., 1960.
[56] Ricou, F. P. and Spalding, D. B., "Measurement of Entrainment by Axisymmetrical Turbulent Jets," *Journal of Fluid Mechanics*, Vol. 11, 1961, pp. 21-32.
[57] Sforza, P., "Mass Momentum and Energy Transport in Turbulent Free Jets," *International Journal of Heat and Mass Transfer*, Vol. 21, 1978, pp. 271-384.
[58] Becker, H. A., Hottel, H. C., and Williams, G. C., "The Nozzle Fluid Concentration Field of the Round Turbulent Free Jet," *Journal of Fluid Mechanics*, Vol. 30, 1967, pp. 285-303.
[59] Way, J. and Libby, P. A., "Application of Hot-Wire Anemometry and Digital Techniques to Measurements in a Turbulent Helium Jet," *AIAA Journal*, Vol. 9, Aug. 1971, pp. 1567-1573.
[60] Birch, A. D., Brown, D. R., Dodson, M. G., and Thomas, J. R., "The Turbulent Concentration Field of a Methane Jet," *Journal of Fluid Mechanics*, Vol. 88, Pt. 3, 1978, pp. 431-449.
[61] Antonia, R. A., Prabhu, A., and Stephenson, S. F., "Conditionally Sampled Measurements in a Heated Turbulent Jet," *Journal of Fluid Mechanics*, Vol. 72, 1975, pp. 455-480.
[62] Brown, G. and Roshko, A., "The Effect of Density Difference on the Turbulent Mixing Layers," *Turbulent Shear Flows*, AGARD CP-93, 1971.
[63] Waltrup, P. J. and Schetz, J. A., "Tangential Slot Injection of Carbon Dioxide and Helium into a Supersonic Air Stream," American Society of Mechanical Engineers, Paper 72-WA/FE-37, Nov. 1972.
[64] Winant, C. D. and Browand, F. K., "Vortex Pairing: The Mechanism of Turbulent Mixing Layer Growth at Moderate Reynolds Numbers," *Journal of Fluid Mechanics*, Vol. 63, Pt. 2, 1974, pp. 237-255.
[65] Chandrsuda, C. and Bradshaw, P., "An Assessment of the Evidence for Orderly Structure in Turbulent Mixing Layers," Imperial College, London, Aero. Rept. 75-03, 1975.
[66] Lau, J. C. and Fisher, M. J., "The Vortex-Street Structure of Turbulent Jets," *Journal of Fluid Mechanics*, Vol. 67, Pt. 2, 1975, pp. 299-337.
[67] Yule, A. J., "Observations of Late Transitional and Turbulent Flow in Round Jets," *Turbulent Shear Flows I*, edited by F. Durst, B. E. Launder, F. W. Schmidt, and J. H. Whitelaw, Springer-Verlag, Berlin, 1979.
[68] Tollmien, W., "Berechnung Turbulenter Ausbreitungsvorgänge," *Zeitschrift für Angewandte Mathematik und Mechanik,* Vol. 6, 1926, pp. 468-478.
[69] Görtler, H., "Berechnung von Aufgaben der Freien Turbulenz auf Grund eines Neuen Näherungsansatzes," *Zeitschrift für Angewandte Mathematik und Mechanik*, Vol. 22, 1942, pp. 244-254.
[70] Kotsovinos, N. E., "A Note on the Conservation of the Axial Momentum of a Turbulent Jet," *Journal of Fluid Mechanics*, Vol. 87, Pt. 1, 1978, pp. 55-63.
[71] Schetz, J. A. and Jannone, J., "A Study of Linearized Approximations to the Boundary Layer Equations," *Journal of Applied Mechanics*, Dec. 1965, pp. 757-764.
[72] Clauser, F. H., "The Turbulent Boundary Layer," *Advances in Applied Mechanics*, Vol. IV, Pergamon Press, London, 1956.
[73] Ting, L. and Libby, P. A., "Remarks on the Eddy Viscosity in Compressible Mixing Flows," *Journal of the Aerospace Sciences*, Vol. 27, Oct. 1960, pp. 797-798.
[74] Ferri, A., Libby, P. A., and Zakkay, V., "Theoretical and Experimental Investigation of Supersonic Combustion," *Third ICAS Conference*, Stockholm, 1962.
[75] Bloom, M. H. and Steiger, M. H., "Diffusion and Chemical Reaction in Free Mixing," *IAS Annual Meeting*, Paper 63-67, 1963.
[76] Schetz, J. A., "Supersonic Diffusion Flames," *Supersonic Flow, Chemical Processes and Radiative Transfer*, Pergamon Press, London, 1964.
[77] Schetz, J. A., "Turbulent Mixing of a Jet in a Coflowing Stream," *AIAA Journal*, Vol. 6, Oct. 1968, pp. 2008-2010.
[78] Schlichting, H., "Über das ebene Windschattenproblem," *Ingenieur-Archiv,* Vol. 1, 1930, pp. 533-571.
[79] Van Der Hegge Zijnen, G. G., "Measurements of Turbulence in a Plane Jet of Air by the Diffusion Method and by the Hot-Wire Method," *Applied Scientific Research*, Vol. A7, 1958, pp. 293-312.

[80]Cooper, R. D. and Lutzky, M., "Exploratory Investigation of the Turbulent Wakes Behind Bluff Bodies," David Taylor Model Basin, DTMB R & D Rept. 953, Oct. 1955.

[81]Launder, B., Morse, A., Rodi, W., and Spalding. D. B., "Prediction of Free Shear Flows—A Comparison of the Performance of Six Turbulence Models," *Free Turbulent Shear Flows*, NASA SP-321, 1973.

[82]Harsha, P. T., "Prediction of Free Turbulent Mixing Using a Turbulent Kinetic Energy Method," *Free Turbulent Shear Flows*, NASA SP-321, 1971.

[83]Rotta, J. C., "Recent Attempts to Develop a Generally Applicable Calculation Method for Turbulent Shear Flow Layers," *Turbulent Shear Flows*, AGARD CP-93, 1971.

[84]Rodi, W., "Basic Equations for Turbulent Flow in Cartesian and Cylindrical Coordinates," Imperial College, BL/TN/A/36, Sept. 1970.

[85]Schetz, J. A., "Free Turbulent Mixing in a Co-flowing Stream," *Free Turbulent Shear Flows,* NASA SP-321, 1971.

[86]*Proceedings of the AFOSR-IFP Stanford Conference on Turbulent Boundary Layer Prediction*, Stanford Univ., 1968.

[87]McDonald, H. and Camarata, F. J., "An Extended Mixing Length Approach for Computing Turbulent Boundary Layer Development," *Proceedings of the AFOSR-IFP Stanford Conference on Turbulent Boundary Layer Prediction,* Stanford Univ., 1968.

[88]Peters, C. and Phares, W. J., "An Integral Turbulent Kinetic-Energy Analysis of Free Shear Flows," *Free Turbulent Shear Flows*, NASA SP-321, 1971.

[89]Spalding, D. B., "The Calculation of the Length Scale of Turbulence in Some Turbulent Boundary Layers Remote from Walls," Imperial College, TWF/TN/31, Sept. 1967.

[90]Rotta, J. C., "Statische Theorie Nichthomogener Turbulenz," *Zeitschrift für Physik,* Vol. 125, 1951, p. 547, and Vol. 131, 1951, p. 51.

[91]Eggers, J. M., "Turbulent Mixing of Coaxial Compressible Hydrogen-Air Jets," NASA TN D-6487, 1971.

[92]Sill, B. and Schetz, J. A., "The Interrelationship Between Eddy Viscosity Mixing Length, Entrainment and Width Growth Models in Turbulent Flows," American Society of Mechanical Engineers, Paper 77-FE-20, June 1977.

[93]List, E. J. and Imberger, J., "Turbulent Entrainment in Buoyant Jets and Plumes," *Journal of the Hydraulics Division, ASCE*, Vol. 99, Sept. 1973, pp. 1461-1474.

[94]Morton, B. R., "On a Momentum-Mass Flux Diagram for Turbulent Jets, Plumes, and Wakes," *Journal of Fluid Mechanics,* Vol. 11, 1961, pp. 21-32.

[95]Morgenthaler, J. A. and Zelazny, S. W., "Predictions of Axisymmetric Free Turbulent Shear Flows Using a Generalized Eddy-Viscosity Approach," *Free Turbulent Shear Flows,* NASA SP-321, 1971.

[96]Rotta, J. C., "Inductive Treatment of Prandtl's 1945 Equations for Fully Developed Turbulence," Inst. for Fluid Mechanics, Göttingen, Germany, DLR-FB-74-51, 1974.

[97]Spalding, D. B., "Concentration Fluctuations in a Round Turbulent Free Jet," *Chemical Engineering Science*, Vol. 26, 1971, pp. 95-107.

[98]Hanjalic, K. and Launder, B. E., "A Reynolds Stress Model of Turbulence and Its Application to Thin Shear Flows," *Journal of Fluid Mechanics*, Vol. 52, 1972, p. 609.

[99]Biringen, S., "The Prediction of an Axisymmetric Turbulent Jet by a Three-Equation Model of Turbulence," Von Karman Inst., Rhode-Saint-Genese, Belgium, TN 114, July 1975.

[100]Donaldson, C. DuP., "A Progress Report on an Attempt to Construct an Invariant Model of Turbulent Shear Flows," *Turbulent Shear Flows*, AGARD CP-93, 1971.

[101]Daly, B. J. and Harlow, F. H., "Transport Equations in Turbulence," *The Physics of Fluids*, Vol. 13, 1970, p. 2634.

[102]Cormack, D. E., Leal, L. G., and Seinfeld, J. H., "An Evaluation of Mean Reynolds Stress Turbulence Models: The Triple Velocity Correlation," *Journal of Fluids Engineering*, Vol. 100, 1978, pp. 47-54.

[103]Launder, B. E., Reece, G. J., and Rodi, W., "Progress in the Development of a Reynolds Stress Turbulence Closure," *Journal of Fluid Mechanics*, Vol. 68, Pt. 3, 1975, pp. 537-566.

[104]Launder, B. E. and Morse, A., "Numerical Prediction of Axisymmetric Free Shear Flows with a Second-Order Reynolds Stress Closure," *Turbulent Shear Flows I,* edited by F. Durst, B. E. Launder, F. W. Schmidt, and J. H. Whitelaw, Springer-Verlag, Berlin, 1979.

[105]Brodkey, R. S. (ed.), *Turbulence in Mixing Operations,* Academic Press, New York, 1975.

[106]Helmbold, H. B., Luessen, G., and Heinrich, A. M., "An Experimental Comparison of Constant-Pressure and Constant-Area Jet Pumps," Univ. of Wichita, Engineering Rept. 147, July 1954.

[107]Barchilon, M. and Curtet, R., "Some Details of the Structure of an Axisymmetric Confined Jet with Backflow," *Journal of Basic Engineering*, Dec. 1964, pp. 77-787.

[108]Fekete, G. I., "Two-dimensional Self-preserving Turbulent Jets in Streaming Flow," McGill Univ., Rept. MERL 70-11, 1970.

[109]Oosthuizen, P. H. and Wu, M. C., "Experimental and Numerical Study of Constant Diameter Ducted Jet Mixing," *Turbulent Shear Flows I*, edited by F. Durst, B. E. Launder, F. W. Schmidt, and J. H. Whitelaw, Springer-Verlag, Berlin, 1979.

[110]Curtet, R., "Confined Jets and Recirculation Phenomena with Cold Air," *Combustion and Flame*, Vol. 2, 1958, pp. 383-411.

[111]Ginevskii, A. S., "Turbulent Wake and Jet in a Flow in the Presence of a Longitudinal Pressure Gradient," *Mekhanika i Mashin*, Vol. 2, 1959, pp. 31-36.

[112]Curtet, R. and Ricou, F. P., "On the Tendency to Self-Preservation in Axisymmetric Ducted Jets," *Journal of Basic Engineering*, Dec. 1964, pp. 765-776.

[113]Hill, P. G., "Turbulent Jets in Ducted Streams," *Journal of Fluid Mechanics*, Vol. 22, Pt. 1, 1965, pp. 161-186.

[114]Schetz, J. A. and Jannone, J., "Planar Free Turbulent Mixing with an Axial Pressure Gradient," *Journal of Basic Engineering*, Vol. 33, Dec. 1967, pp. 707-714.

[115]Tennankore, K. N. and Steward, F. R., "Comparison of Several Turbulence Models for Predictions Flow Patterns within Confined Jets," *Turbulent Shear Flows I*, edited by F. Durst, B. E. Launder, F. W. Schmidt, and J. H. Whitelaw, Springer-Verlag, Berlin, 1979.

[116]Hedges, K. R. and Hill, P. G., "Compressible Flow Ejectors Part II—Flow Field Measurment and Analysis," *Journal of Fluids Engineering*, Vol. 96, Sept. 1974, pp. 282-288.

[117]Ridjanovic, M., "Wake with Zero Change of Momentum Flux," Ph.D. Dissertation, Univ. of Iowa, 1963.

[118]Wang, H., "Flow Behind a Point Source of Turbulence," Ph.D. Dissertation, Univ. of Iowa, 1963.

[119]Naudascher, E., "Flow in the Wake of Self-Propelled Bodies and Related Sources of Turbulence," *Journal of Fluid Mechanics*, Vol. 22, Pt. 4, 1965, pp. 625-656.

[120]Ginevskii, A. S., Pochkina, K. A., and Ukhanova, L. N., "Propagation of Turbulent Jet Flow with Zero Excess Impulse," *Mekhanika Zhidkosti i Gaza*, Vol. 1, June 1966, pp. 164-166.

[121]Schetz, J. A. and Jakubowski, A. K., "Experimental Studies of the Turbulent Wake Behind Self-Propelled Slender Bodies," *AIAA Journal*, Vol. 13, Dec. 1975, pp. 1568-1575.

[122]Schetz, J. A. and Favin, S., "Analysis of Free Turbulent Mixing Flows Without a Net Momentum Defect," *AIAA Journal*, Vol. 10, Nov. 1972, pp. 1524-1526.

[123]Swanson, R. C., Jr. and Schetz, J. A., "Calculations of the Turbulent Wake Behind Slender Self-Propelled Bodies with a Kinetic Energy Method," *Journal of Hydronautics*, Vol. 9, April 1975, pp. 78-80.

[124]Schetz, J. A. and Favin, S., "Numerical Solution for the Near Wake of a Body with Propeller," *Journal of Hydronautics*, Vol. 11, Oct. 1977, pp. 136-141.

[125]Schetz, J. A. and Favin, S., "Numerical Solution of a Body-Propeller Combination including Swirl and Comparisons with Data," *Journal of Hydronautics*, Vol. 13, April 1979, pp. 46-51.

[126]Lewellen, W. S., Teske, M., and Donaldson, C. DuP., "Turbulent Wakes in a Stratified Fluid," Aeronautical Research Associates of Princeton, Rept. 226, Aug. 1974.

[127]Rose, W. G., "A Swirling Round Turbulent Jet," *Journal of Applied Mechanics*, 1962, pp. 615-625.

[128]Chigier, N. A. and Chervinsky, A., "Experimental Investigation of Swirling Vortex Motion in Jets," *Journal of Applied Mechanics*, 1967, pp. 443-451.

[129]Pratte, B. D. and Keffer, J. F., "The Swirling Turbulent Jet," *Journal of Basic Engineering*, Vol. 95, 1973.

[130]Grandmaison, E. W. and Becker, H. A., "Turbulent Mixing in Free Swirling Jets," *Turbulent Shear Flows I*, edited by F. Durst, B. E. Launder, F. W. Schmidt, and J. H. Whitelaw, Springer-Verlag, Berlin, 1979.

[131]Schetz, J. A. and Swanson, R. C., Jr., "Turbulent Jet Mixing at High Supersonic Speeds," *Zeitschrift für Flugwissenschaften*, Vol. 21, May 1973.

[132]Schetz, J. A., Daffan, E. B., and Jakubowski, A. K., "Turbulent Wake Behind Slender Propeller-Driven Bodies at Angle of Attack," *AIAA Journal,* Vol. 16, Jan. 1978, pp. 6-8.

[133]Swean, T. F. and Schetz, J. A., "Flow About a Slender Propeller-Driven Body in Temperature Stratified Fluid," *AIAA Journal*, Vol. 17, Aug. 1979, pp. 863-869.

[134]Görtler, H., "Decay of Swirl in an Axially Symmetric Jet Far from the Orifice," *Revista Matematica Hispano Americana,* 4ª Serie, Tomo IV, Nos. 4 and 5, 1954.

[135]Lilley, D. G., "Prediction of Inert Turbulent Swirl Flows," *AIAA Journal*, Vol. 11, July 1973, pp. 955-960.

[136]Rodi, W. and Spalding, D. B., "A Two-Parameter Model of Turbulence and Its Application to Free Jets," *Wärme und Stoffübertragung,* Vol. 3, 1970, pp. 585-595.

[137]Rodi, W., private communication, 1977.

[138]Lumley, J. L., "Prediction Methods for Turbulent Flow," Von Karman Inst., Rhode-Saint-Genese, Belgium, Lecture Ser. 76, 1975.

[139]Soo, S. L., *Fluid Dynamics of Multiphase Systems*, Blaisdell Publishing Co., Waltham, Mass., 1967.

[140]Goldschmidt, V. W., Householder, M. K., Ahmadi, G., and Chuang, S. C., "Turbulent Diffusion of Small Particles Suspended in Turbulent Jets," *Progress in Heat and Mass Transfer*, Vol. 6, Pergamon Press, Oxford, England, 1972.

[141]Lilly, G. P., "Effect of Particle Size on Eddy Diffusivity," *Industrial Engineering Chemistry Fundamentals*, Vol. 12, March 1973, pp. 268-275.

[142]Hedman, P. O. and Smoot, L. D., "Particle-Gas Dispersion Effects in Confined Coaxial Jets," *AIChE Journal*, Vol. 21, Feb. 1975, pp. 372-379.

[143]Laats, M. K., "Experimental Study of the Dynamics of an Air-Dust Jet," *Inzhenerno-Fizicheskii Zhurnal*, Vol. 10, Jan. 1966.

[144]Laats, M. K. and Frishman, F. A., "Scattering of an Inert Admixture of Different Grain Size in a Two-Phase Axisymmetric Jet," *Heat Transfer—Soviet Research*, Vol. 2, 1970, pp. 7-11.

[145]Laats, M. K. and Frishman, M. K., "Assumptions Used in Calculating the Two-Phase Jet," *Mekhanica Zhidkosti i Gaza,* Vol. 5, Feb. 1970, pp. 186-191.

[146]Melville, W. K. and Bray, K. N. C., "The Two-Phase Turbulent Jet," *International Journal of Heat and Mass Transfer*, Vol. 22, 1979, pp. 279-287.

[147]Hinze, J. O., "Turbulent Fluid and Particle Interaction," *Progress in Heat and Mass Transfer*, Vol. 6, edited by G. Hetsroni, S. Sideman, and J. P. Hartnett, Pergamon Press, Oxford, 1972.

[148]Abramovich, G. N., "The Effect of Admixture of Solid Particles on the Structure of a Turbulent Gas Jet," *Soviet Physics—Doklady, Fluid Mechanics,* Vol. 15, Aug. 1970.

[149]Abramovich, G. N., Bazhanov, V. I., and Girshovich, T. A., "A Turbulent Jet with Heavy Components," *Mekhanika Zhidkosti i Gaza,* Nov.-Dec. 1972, pp. 41-49.

[150]Vasil'kov, A. P., "Calculation of a Two-Phase Isobaric Jet," *Mekhanika Zhidkosti i Gaza*, Sept.-Oct. 1976, pp. 57-63.

[151]Sforza, P. M., Steiger, M. H., and Trentacoste, N., "Studies on Three-Dimensional Viscous Jets," *AIAA Journal*, Vol. 4, May 1966, pp. 800-806.

[152]Yevdjevich, V. M., "Diffusion of Slot-Jets with Finite Orifice Length-Width Ratios," Colorado State Univ., Fort Collins, Colo., Hydraulic Paper 2, 1966.

[153]Trentacoste, N. and Sforza, P. M., "Further Experimental Results for Three-Dimensional Free Jets," *AIAA Journal*, Vol. 5, May 1967, pp. 855-891.

[154]Sfeier, A. A., "The Velocity and Temperature Fields of Rectangular Jets," *International Journal of Heat and Mass Transfer*, Vol. 19, 1976.

[155]Lorber, A. K. and Schetz, J. A., "Turbulent Mixing of Multiple Co-Axial Gaseous Fuel Jets in a Supersonic Airstream," *AIAA Journal*, Vol. 13, Aug., 1975, pp. 973-974.

[156]DeJoode, A. D. and Patanker, S. V., "Prediction of Three-Dimensional Turbulent Mixing in an Ejector," *AIAA Journal*, Vol. 16, Feb. 1978, pp. 145-150.

[157]McGuirk, J. J. and Rodi, W., "The Calculation of Three-Dimensional Turbulent Jets," *Turbulent Shear Flows I,* edited by F. Durst, B. E. Launder, F. W. Schmidt, and J. H. Whitelaw, Springer-Verlag, Berlin, 1979.

[158]Campbell, J. F. and Schetz, J. A., "Analysis of Injection of a Heated Turbulent Jet into a Cross Flow," NASA TR R-413, Dec. 1973.

[159]Jordinson, R., "Flow in a Jet Directed Normal to the Wind," British Aeronautical Research Council, R & M 3074, Oct. 1956.

[160] Keffer, J. F. and Baines, W. D., "The Round Turbulent Jet in a Cross-Wind," *Journal of Fluid Mechanics*, Vol. 15, Pt. 4, 1963, pp. 481-496.

[161] Kamotani, Y. and Greber, I., "Experiments on a Turbulent Jet in a Cross-Flow," Case Western Reserve Univ., FTAS/TR-71-62, June 1971; also *AIAA Journal,* Vol. 10, Nov. 1972, pp. 1425-1429.

[162] Pratte, B. D. and Baines, W. D., "Profiles of the Round Turbulent Jet in a Cross Flow," *Proceedings of ASCE, Journal of the Hydraulics Division,* Nov. 1967, pp. 56-63.

[163] McMahon, H. M. and Antani, D. L., "An Experimental Study of a Jet Issuing from a Lifting Wing," *Journal of Aircraft*, Vol. 16, April 1979, pp. 275-281.

[164] Chassaing, P., George, J., Claria, A., and Sananes, F, "Physical Characteristics of Subsonic Jets in a Cross-Stream," *Journal of Fluid Mechanics*, Vol. 62, Pt. 1, 1974, pp. 41-64.

[165] Platten, J. L. and Keffer, J. F., "Deflected Turbulent Jet Flows," *Journal of Applied Mechanics*, 1971, pp. 756-758.

[166] Krausche, D., Fearn, R. L., and Weston, R. P., "Round Jet in a Cross Flow: Influence of Injection Angle on Vortex Properties," *AIAA Journal*, Vol. 16, June 1978, pp. 636-637.

[167] Salzman, R. N. and Schwartz, S. H., "Experimental Study of a Solid-Gas Jet Issuing into a Transverse Stream," *Journal of Fluids Engineering*, Vol. 100, Sept. 1978, pp. 333-339.

[168] Rudinger, G., "Some Aspects of Gas-Particle Jets in a Cross Flow," American Society of Mechanical Engineers, Paper 75-WA/HT-5, 1975.

[169] Smoot, L. D. and Allred, L. D., "Particle-Gas Mixing in Confined Nonparallel Coaxial Jets," *AIAA Journal*, Vol. 13, June 1975, pp. 721-722.

[170] Morkovin, M. V., Pierce, C. A., Jr., and Craven, C. E., "Interaction of a Side Jet with a Supersonic Main Stream," Univ. of Michigan, Engineering Research Bull. 35, Sept. 1952.

[171] Schetz, J. A., Hawkins, P. F., and Lehman, H., "The Structure of Highly Underexpanded Transverse Jets in a Supersonic Stream," *AIAA Journal*, Vol. 5, May 1967, pp. 882-884.

[172] Schetz, J. A., Weinraub, R., and Mahaffey, R., "Supersonic Transverse Jets in a Supersonic Stream," *AIAA Journal*, Vol. 6, May 1968, pp. 933-934.

[173] Orth, R. C., Schetz, J. A., and Billig, F. S., "The Interaction and Penetration of Gaseous Jets in Supersonic Flow," NASA CR-1386, July 1969.

[174] Schetz, J. A., "Interaction Shock Shape for Transverse Injection," *Journal of Spacecraft and Rockets*, Vol. 7, Feb. 1970, pp. 143-149.

[175] Dowdy, M. W. and J. F. Newton, "Investigation of Liquid and Gaseous Secondary Injection Phenomena on a Flat Plate with $M=2.01$ to $M=4.54$," Jet Propulsion Lab., TR 32-542, 1963.

[176] Kolpin, M., Horn, K., and Reichenbach, R., "Study of Penetration of a Liquid Injectant into a Supersonic Flow," *AIAA Journal*, Vol. 6, May 1968, pp. 853-858.

[177] Sherman, A. and Schetz, J. A., "Breakup of Liquid Sheets and Jets in a Supersonic Gas Stream," *AIAA Journal*, Vol. 9, April 1971, pp. 666-673.

[178] Kush, E. A., Jr. and Schetz, J. A., "Liquid Jet Injection into a Supersonic Flow," *AIAA Journal*, Vol. 11, Sept. 1973, pp. 1223-1224.

[179] Joshi, P. B. and Schetz, J. A., "Effect of Injector Geometry on the Structure of a Liquid Jet Injected Normal to a Supersonic Airstream," *AIAA Journal*, Vol. 13, Sept. 1975, pp. 1137-1138.

[180] Yates, C. L. and Rice, J. L. "Liquid Jet Penetration," Applied Physics Lab., Johns Hopkins Univ., Research and Development Programs Quarterly Rept. U-RQR/69-2, 1969.

[181] Baranovsky, S. I. and Schetz, J. A., "Effect of Injection Angle on Liquid Injection in Supersonic Flow," AIAA Paper 79-0383, 1979.

[182] Schetz, J. A. and Padhye, A., "Penetration and Break up of Liquids in Subsonic Airstreams," *AIAA Journal*, Vol. 15, Oct. 1977, pp. 1390-1395.

[183] Schetz, J. A. and Billig, F. S., "Penetration of Gaseous Jets Injected into a Supersonic Stream," *Journal of Spacecraft and Rockets*, Vol. 3, Nov. 1966, pp. 1658-1665.

[184] Billig, F. S., Orth, R. C., and Lasky, M., "A Unified Analysis of Gaseous Jet Penetration," *AIAA Journal*, Vol. 9, June 1971, pp. 1048-1058.

[185] Thomson, J. F., "Two Approaches to the Three-Dimensional Jet-in-Cross-Wind Problem: A Vortex Lattice Model," Ph.D. Thesis, Georgia Inst. of Technology, 1971.

[186] Chien, C. J. and Schetz, J. A., "Numerical Solution of the Three-Dimensional Navier-Stokes Equations with Application to Channel Flows and a Buoyant Jet in a Cross-Flow," *Journal of Applied Mechanics*, Vol. 42, March 1975, pp. 575-579.

[187] Chen, C. J. and Rodi, W., "A Review of Experimental Data of Vertical Turbulent Buoyant Jets," *The Science and Applications of Heat and Mass Transfer*, Vol. II, edited by D. B. Spalding, Pergamon Press, Oxford, 1978.

[188]Harris, P. R., "The Densimetric Flows Caused by the Discharge of Heated Two-Dimensional Jets Beneath a Free Surface," Ph.D. Thesis, Univ. of Bristol, Dept. of Civil Engineering, 1967.

[189]Kotsovinos, N. E., "A Study of the Entrainment and Turbulence in a Plane Buoyant Jet," W. M. Keck Lab. of Hydraulics and Water Resources, California Inst. of Technology, Pasadena, Calif., Rept. KH-R-32, 1975.

[190]Rouse, H., Yih, C. S., and Humphreys, H. W., "Gravitational Convection from a Boundary Source," *Tellus*, Vol. 4, 1952, pp. 201-210.

[191]Pryputniewicz, R. J. and Bowley, W. W., "A Experimental Study of Vertical Buoyant Jets Discharged into Water of Finite Depth," *Journal of Heat Transfer,* May 1975, pp. 274-281.

[192]Cleeves, V. and Boelter, L. M. K., "Isothermal and Non-Isothermal Air-Jet Investigations," *Chemical Engineering Progress,* Vol. 43, March 1947, pp. 123-134.

[193]Hayashi, T. and Ito, M., "Initial Dilution of Effluent Discharging into Stagnant Sea Water," *International Symposium Disharge of Sewage from Sea Outfalls,* Paper 26, London, Sept. 1974.

[194]Abraham, G., "Jet Diffusion in a Liquid of Greater Density," *ASCE Journal of the Hydraulics Division*, Vol. HY6, No. 86, 1960, pp. 1-13.

[195]Turner, J. S., "Jets and Plumes with Negative or Reversing Buoyancy," *Journal of Fluid Mechanics,* Vol. 26, 1966, pp. 792-799.

[196]Crawford, T. V. and Leonard, A. S., "Observations of Buoyant Plumes in Calm Stably Stratified Air," *Journal of Applied Meteorology*, Vol. 1, 1962, pp. 251-256.

[197]Morton, B. R., Taylor, G. I., and Tumer, J. S., "Turbulent Gravitational Convection from Maintained and Instantaneous Sources," *Proceedings of the Royal Society* (*London*), Vol. A234, 1956, pp. 1-23.

[198]Fox, D. G., "Forced Plume in a Stratified Fluid," *Journal of Geophysical Research,* Vol. 75, No. 33, 1970, pp. 6818-6835.

[199]Fan, L. N., "Turbulent Buoyant Jets into Stratified or Flowing Ambient Fluids," Div. of Engineering and Allied Science, California Inst. of Technology, Pasadena, Calif., Rept. KH-R-15, 1967.

[200]Sneck, H. J. and Brown, D. M., "Plume Rising from Large Thermal Sources Such as Dry Cooling Towers," *Journal of Heat Transfer,* 1974, pp. 232-238.

[201]Abraham, G. and Eysink, W. D., "Jets Issuing into Fluid with a Density Gradient," Delft Hydronautics Lab., Publ. 66, 1969.

[202]George, W. K., Alpert, R. L., and Tamanini, F., "Turbulence Measurements in an Axisymmetric Plume," Factory Mutual, Research Rept. 22359-2, 1976.

[203]Oosthuizen, P. H., "Profile Measurements in Vertical Axisymmetric Buoyant Air Jets," *6th International Heat Transfer Conference,* Paper MC-18, Toronto, Canada, 1978.

[204]Merritt, G. E., "Wake Growth and Collapse in Stratified Flow," *AIAA Journal,* Vol. 12, July 1974, pp. 940-949.

[205]Schooley, A. H. and Stewart, R. W., "Experiments with a Self-Propelled Body Submerged in a Fluid with Vertical Density Gradient," *Journal of Fluid Mechanics*, Vol. 15, 1963, pp. 83-96.

[206]Schooley, A. H., "Wake Collapse in a Stratified Fluid," *Science*, July 1967.

[207]Sundaram, T. R., Straton, J. E., and Rehm, R. C., "Turbulent Wakes in a Stratified Medium," Cornell Aeronautical Lab., Rept. AG-3018-A-1, Nov. 1971.

[208]Van der Watering, W. P. M., Tulin, M. P., and Wu, J., "Experiments on Turbulent Wakes in a Stable Density-Stratified Environment," Hydronautics, Inc., TR 231-24, Feb. 1969.

[209]Wright, S. J., "An Entrainment Model for Buoyant Jet Discharges," *Proceedings of the Heat Transfer and Fluid Mechanics Institute*, 1978.

[210]Hwang, S. S. and Pletcher, R. H., "Prediction of Buoyant Turbulent Jets and Plumes in a Cross Flow," *6th International Heat Transfer Conference*, Paper MC-19, Toronto, Canada, 1978.

[211]Ayoub, G. M., "Dispersion of Buoyant Jets in a Coflowing Ambient Flow," Ph.D. Thesis, Univ. of London, 1971.

[212]Hossain, M. S. and Rodi, W., "Influence of Buoyancy on the Turbulence Intensities in Horizontal and Vertical Jets," *6th ICHMT Seminar of Turbulent Buoyant Convention*, Dubrovnik, Yugoslavia, 1976.

[213]Gibson, M. M. and Launder, B. E., "On the Calculation of Horizontal Non-Equilibrium Turbulent Shear Flows Under Gravitational Influence," *Journal of Heat Transfer,* Feb. 1976, pp. 81-87.

[214]Chen, C. J. and Rodi, W., "A Mathematical Model for Stratified Turbulent Flows and Its Application to Buoyant Jets," *XVIth Congress, Institute of the Association of Hydronautics Research,* Sao Paolo, Brazil, 1975.

[215]Chen, C. J. and Chen, C. H., "On Prediction and Unified Correlation for Decay of Vertical Buoyant Jets," American Society of Mechanical Engineers, Paper 78-HT-21, 1978.

[216]McGuirk, J. J. and Rodi, W., "Calculations of Three Dimensional Heated Surface Jets," *9th ICHMT International Conference on Turbulent Buoyant Convection*, Dubrovnik, Yugoslavia, Aug. 1976.

[217]Kovasznay, L. S. G. and Ali, S. F., "Structure of the Turbulence in the Wake of a Heated Flat Plate," *Proceedings of the 5th International Heat Transfer Conference*, Tokyo, 1975.

[218]Chevray, R. and Kovasznay, L. S. G., "Turbulence Measurements in the Wake of a Thin Flat Plate," *AIAA Journal*, Vol. 7, 1969, p. 8.

[219]Hassid, S., "Collapse of Turbulent Wakes in Stably Stratified Media," Pennsylvania State Univ., TM-77-237, Aug. 1977.

[220]Lin, J. T. and Pao, Y. H., "Turbulent Wake of a Self-Propelled Slender Body in Stratified and Nonstratified Fluids: Analysis and Flow Visualization," Flow Research Inc., Rept. 11, July 1973.

[221]Lin, J. T. and Pao, Y. H., "Velocity and Density Measurements in the Turbulent Wake of a Propeller-Driven Slender Body in a Stratified Fluid," Flow Research Inc., Rept. 36, Aug. 1974.

[222]Pao, Y. H. and Lin, J. T., "Turbulent Wake of a Towed Slender Body in Stratified and Nonstratified Fluids: Analysis and Flow Visualizations," Flow Research Inc., Rept. 10, June 1973.

[223]Pao, Y. H. and Lin, J. T., "Velocity and Density Measurements in the Turbulent Wake of a Towed Slender Body in Stratified and Nonstratified Fluids," Flow Research Inc., Rept. 12, Dec. 1973.

[224]Lewellen, W. S., Teske, M. E., and Donaldson, C. DuP., "Examples of Variable Density Flows Computed by a Second-Order Closure Description of Turbulence," *AIAA Journal*, Vol. 14, March 1976, pp. 382-387.

[225]Murthy, S. N. B. (ed.), *Progress in Astronautics and Aeronautics:* Aerodynamics of Base Combustion, Vol. 40, AIAA, New York, 1976.

[226]Crocco, L. and Lees, L., "A Mixing Theory for the Interaction Between Dissipative Flows and Nearly Isentropic Streams," *Journal of the Aeronautical Sciences,* Vol. 19, Oct. 1952, pp. 649-676.

[227]Alber, I. E., "Integral Theory for Turbulent Base Flows at Subsonic and Supersonic Speeds," Ph. D. Thesis, California Inst. of Technology, Pasadena, Calif., June 1967.

[228]Alber, I. E. and Lees, L., "Integral Theory for Supersonic Turbulent Base Flows," *AIAA Journal*, Vol. 6, July 1968, pp. 1343-1351.

[229]Gilreath, H. E. and Schetz, J. A., "Transition and Mixing in the Shear Layer Produced by Tangential Injection in Supersonic Flow," *Journal of Basic Engineering,* Dec. 1971, pp. 610-618.

[230]Schetz, J. A., Billig, F. S., and Favin, S., "Simplified Analysis of Supersonic Base Flows Including Injection and Combustion," *AIAA Journal*, Vol. 14, Jan. 1976, pp. 7-8.

[231]Kenworthy, M. A. and Schetz, J. A., "Experimental Study of Slot Injection into a Supersonic Stream," *AIAA Journal*, Vol. 11, May 1973, pp. 585-586.

[232]Schetz, J. A., Billig, F. S., and Favin, S., "Approximate Analysis of Axisymmetric Base Flows with Injection in Supersonic Flow," *AIAA Journal* (submitted for publication).

[233]Reid, J. and Hastings, R. C., "Experiments on the Axisymmetric Flow over Afterbodies and Bases at $M = 2.0$," British Aeronautical Research Council ARC 21, 707, Oct. 1959.

[234]Neale, D. H., Hubbartt, J. E., and Strahle, W. C., "Effects of Axial and Radial Air Injection on the Near Wake with and without External Compression," *AIAA Journal*, Vol. 17, March 1979, pp. 301-303.

[235]Pope, A. and Harper, J. J., *Low-Speed Wind Tunnel Testing*, Wiley, New York, 1966.

[236]Pope, A. and Goin, K. L., *High-Speed Tunnel Testing*, Wiley, New York, 1965.

[237]Faro, I. D. V. (ed.), *Handbook of Supersonic Aerodynamics, Wind Tunnel Instrumentation and Operation*, Vol. 6. 20, U. S. Government Printing Office, Washington, D.C., 1961.

[238]Chue, S. H., "Pressure Probes for Fluid Measurements," *Progress in Aerospace Science,* Vol. 16, Feb. 1975, pp. 147-223.

[239]Benedict, R. P., *Fundamentals of Temperature, Pressure and Flow Measurements,* Wiley, New York, 1977.

[240]Ladenburg, R. W. (ed. Pt. 1), Lewis, B., Pease, R. N., and Taylor, H. S., "Physical Measurements in Gas Dynamics and Combustion," *High Speed Aerodynamics and Jet Propulsion,* Vol. IX, Princeton University Press, Princeton, N. J., 1954.

[241]Richards, B. E. (ed.), *Measurement of Unsteady Fluid Dynamic Phenomena,* Hemisphere Publishing Corp., Washington, D.C., 1976.

[242]Cooper, R. D. and Tulin, M. P., *Turbulence Measurements with the Hot-Wire Anemometer,* AGARDograph 12, 1955.

[243]Bradshaw, P., *An Introduction to Turbulence and Its Measurement*, The Commonwealth and International Library, Pergamon Press, Oxford, 1971.

[244]Durst, F., Melling, A., and Whitelaw, J. H., *Principles and Practice of Laser-Doppler Anemometry,* Academic Press, London, 1976.

[245]Watrasiewicz, B. M. and Rudd, M. J., *Laser Doppler Measurements,* Butterworths, London, 1976.

[246]Duranni, T. S. and Greated, C. A., *Laser Systems in Flow Measurements,* Plenum Press, New York, 1977.